# FISH
# Handling and Processing

# FISH
# Handling and Processing

– Author –
*Mehr Deeba*

Wordpen Academics
Publishing Science That Matter

**ISBN: 978-93-90438-45-7 (Hardback)**

**Published by :** **WORDPEN ACADEMICS**
**4736/23, Ansari Road, Darya Ganj,**
**New Delhi – 110 002**
**Phone: +91-011-35004336**
**E-mail: info@wordpenacademics.com**

***Printed at*** **:** **Replika Press Pvt. Ltd.**

*Exclusive Distributor:* ***Bio-Green Books, New Delhi***

# Preface

The term fish processing refers to the processes associated with fish and fish products between the time fish are caught or harvested, and the time the final product is delivered to the customer. Although the term refers specifically to fish, in practice it is extended to cover any aquatic organisms harvested for commercial purposes, whether caught in wild fisheries or harvested from aquaculture or fish farming. Larger fish processing companies often operate their own fishing fleets or farming operations. The products of the fish industry are usually sold to grocery chains or to intermediaries. Fish are highly perishable. A central concern of fish processing is to prevent fish from deteriorating, and this remains an underlying concern during other processing operations. Fish processing can be subdivided into fish handling, which is the preliminary processing of raw fish, and the manufacture of fish products. Another natural subdivision is into primary processing involved in the filleting and freezing of fresh fish for onward distribution to fresh fish retail and catering outlets, and the secondary processing that produces chilled, frozen and canned products for the retail and catering trades. There is evidence humans have been processing fish since the early Holocene. These days, fish processing is undertaken by artisan fishermen, on board fishing or fish processing vessels, and at fish processing plants. An alternative, and obvious way of keeping fish fresh is to keep them alive until they are delivered to the buyer or ready to be eaten. This is a common practice worldwide. Typically, the fish are placed in a container with clean water, and dead, damaged or sick fish are removed. The water temperature is then lowered and the fish are starved to reduce their metabolic rate. This decreases fouling of water with metabolic products (ammonia, nitrite and carbon dioxide) that become toxic and make it difficult for the fish to extract oxygen.

Fish can be kept alive in floating cages, wells and fish ponds. In aquaculture, holding basins are used where the water is continuously filtered and its temperature and oxygen level are controlled. In China, floating cages are constructed in rivers out of palm woven baskets, while in South America simple fish yards are built in the backwaters of rivers. Live fish can be transported by methods which range from simple artisanal methods where fish are placed in plastic bags with an oxygenated atmosphere, to sophisticated systems which use trucks that filter and recycle the water, and add oxygen and regulate temperature. Preservation techniques are needed to prevent fish spoilage and lengthen shelf life. They are designed to inhibit the activity of spoilage bacteria and the metabolic changes that result in the loss of fish quality. Spoilage bacteria are the specific bacteria that produce the unpleasant odours and flavours associated with spoiled fish. Forage fish, which are often caught in large numbers, are usually chilled with refrigerated or chilled seawater. Once chilled or frozen, the fish need further cooling to maintain the low temperature. There are key issues with fish cold store design and management, such as how large and energy efficient they are, and the way they are insulated and palletized. An effective method of preserving the freshness of fish is to chill with ice by distributing ice uniformly around the fish.

It is a safe cooling method that keeps the fish moist and in an easily stored form suitable for transport. It has become widely used since the development of mechanical refrigeration, which makes ice easy and cheap to produce. Ice is produced in various shapes; crushed ice and ice flakes, plates, tubes and blocks are commonly used to cool fish. Particularly effective is slurry ice, made from micro crystals of ice formed and suspended within a solution of water and a freezing point depressant, such as common salt. A more recent development is pumpable ice technology. Pumpable ice flows like water, and because it is homogeneous, it cools fish faster than fresh water solid ice methods and eliminates freeze burns. It complies with HACCP and ISO food safety and public health standards, and uses less energy than conventional fresh water solid ice technologies.

It is expected that the book will provide a fund of rich experiences to the students and teachers.

—*Author*

# Contents

# 1

# Fish Breeding

When Charles Darwin laid down the Theory of the Survival of the Fittest, he was only stating what was so very obvious in nature. The best male of the species will try to impregnate the females, and the females prefer the best male since his DNA increases the chances offspring survival. That is why many animals like the lions etc will have only one dominant male who has successfully defeated all other males in the group. This male alone impregnates the sexually active females. He continues to be the leader until a healthier and stronger male defeat him.

There is something very vital in this theory for the aquarist too. When your fish are ready to breed, it will not do to select just any pair. You need to select the male and female very carefully, unless you are breeding to provide food for other fish. Here are some pointers that will help you choose the right pair:

- Fish that have brighter colours and good markings are the healthiest in a stock. Fish that have good finage, movements and vitality are also healthy.
- The appetite of a fish is often an indicator of its general health and vitality. Only fish that have a good appetite should be selected for breeding.
- Do not use fish that appear stunted or deformed in any way.
- The age of the fish is also important while you consider mating partners. It is better to go for younger fish that are in their prime rather than choosing older fish that have almost reached

the end of the reproductive lives. An exception from this rule can be if you have an older specimen with exceptional code colouration, finage or other qualities that you want to breed on.

- Compatible pairs make good partners. For instance, in Cichlids, fish form pairs only after they have been put together in an aquarium for many months. If the fish are not compatible, one of them may even bully the other to death.
- Unhealthy fish and fish that are not mature enough will not produce healthy fry. You should therefore only choose healthy and mature fish for breeding.
- In case you want to experiment crossing two different strains, please keep in mind that the resulting fry may be unattractive. Never sell crossbreed fish as pure fish. Do not mix impure strains with pure strains.
- Hybrids made by crossing two different species are often sterile. It is possible to mistake the females of different species, because some of them look very similar. Hybrids should be avoided and should never be sold as anything other than hybrids.

## Fish Breeding: Condition the Parents

After choosing the parents, you need to condition the fish so that they spawn. One of the most effective ways to do this is to simulate their natural habitats as closely as possible. Making certain changes in the environment will also help you to encourage the fish to spawn.

Providing the right water conditions is one of the first steps. Conditions that are as similar to their natural habitats should be provided. Proper aquarium set-up that provides for enough "security", hiding places and crevices is essential. The water current, lighting and temperature should also be taken into consideration. Some fish mate only when they are in schools. Isolating these fish will discourage spawning.

Providing the right food is the next important thing to do. The mates that have been selected for spawning should be given food that is very high in protein. Live food seems to be a catalyst in this process. You can feed the fish up to three times a day. Of course, remember to clean up the aquarium after each feeding.

Some fish spawn during the rainy season. It can therefore be a good idea to simulate these conditions in the aquarium too. This is actually a very simple procedure. Reduce the water level in the aquarium to half the normal height. Add 5% of the aquarium volume everyday. Add soft water that is slightly cooler that the aquarium water. A drip system or a spray bar will help to add to the notion. During this time, fish should be fed heavily. This process can be repeated till fish show signs of spawning. Most fish respond to the first change in water conditions - like soft water or altered water temperature.

## Equipment and Breeding Fish

The first and foremost need for successful spawning of many species is space. Fish often become territorial when their breeding time approaches. Most males are very aggressive during this time, and will chase away any potential competitors.

This becomes especially dangerous for the younger fish, and more so if they are unable to get to a safer place, out of reach of the aggressive male. The primary need of fish that are going to breed is enough space for them to chase, hide and move. There are species that needs very little space and that don't get especially aggressive, such as the very popular Guppy.

Some species of fish get into the spawning mode when they are fed highly nutritious food. Thus it is a good practice to give the breeders some special attention. This kind of special attention aims at bringing the fish to their topmost reproductive state. If you have some extra aquariums, place the two sexes separately in the two aquariums.

A delivery aquarium is advisable when your fish are ready to spawn. This will save the fry from being eaten up as food by the other fish. A tight cover should be placed on the aquarium as some fish species can become very nervous during this stage and show a tendency to jump. The conditions in the delivery aquarium should be as close to those in the original aquarium as possible. This will save the parent fish from any undesirable stress at such a critical stage in the breeding cycle.

A special aquarium for rearing fry is also advisable. This can be the same aquarium as the delivery aquarium if the parents are moved from the aquarium once the eggs are laid. If the species guards their young, they should not be removed until they have stopped rearing the fry and started to prepare for a new spawning. The needs and requirements of fry are very different from bigger fish. They need special kinds of food, at least till they grow a bit bigger. Filtration too has to be varied in order to accommodate fry.

As the fry start developing you can perform a 10% water change the first few days, and than raise this up to a 40% water change every day to keep the water clean. Clean water helps keep your fry healthy and fast growing. Optimal survival of the fry is ensured if you segregate the fry depending on their size. The fastest growing fry are almost always the males. The bigger fry will eat the smaller ones if they are not separated.

Some people use a breeding trap as the place where the female livebearer can drop her fry. The fry are thus out of reach as they will be dropped into a separate plastic compartment attached to the rim of the aquarium. If the breeding aquarium is too small to accommodate the female, the chances of her miscarrying while struggling to escape are very high.

The spawning aquariums for different kinds of fish should be set up according to their particular characteristics. Egg scatterers will eat their own eggs. So, care should be taken that their eggs fall away from the reach of the parents. For small species of this kind of fish, a 5-10 gallon aquarium would be the best. The parents should be removed after spawning. If the fish lays non-adhesive eggs, it is better to put nylon netting just above the aquarium floor. The eggs will then fall out of the reach of the parents. In case the eggs are adhesive, then they will be laid among the leaves of plants.

Egg depositors should get aquariums decorated in accordance with their egg depositing behaviour. If they are the types that care for their young, the parents can stay in the aquarium after laying the eggs. It is advisable to provide these fish with broad-leafed plants, flat stones, upturned flowerpots, coconut shells, or rocky caves. If your fish species inhabit rocky areas in the wild, they will appreciate caves or flat stones as breeding sites, and so on. Putting in plants and caves will give fish a sense of security even if they are not plant spawners.

Mouth brooders can be left in the main aquarium even when they are spawning because the eggs as well as the young are well protected by the parent. Brooding does however bring out aggression and brooding fish can cause harm to any other fish in the aquarium. For this reason alone, it may be better to put these fish into a new aquarium while they are breeding.

Setting up a new aquarium for breeding purposes is fairly easy. You do not have to go through all the steps and wait for the Nitrogen cycle, plant your aquarium etc since you can use water and bacteria from your established aquarium. An additional aquarium for breeding purpose need not be very big. Usually an aquarium with a capacity of 2-6 gallons would suffice for most fish.

The best way to set up your aquarium easily is to take water from your existing aquarium and use this to start your new aquarium. Since the fish are being transported from the old aquarium, they will already be acclimatised to this water. Using a foam filter for a few weeks in your old aquarium will help you collect a good quantity of beneficial bacteria that can be easily moved to the new aquarium. Decorations and plants can be kept to a minimum for most species. Changing water in this aquarium regularly is a must.

The best way to clean out the material wastes in this aquarium is to siphon it out using a flexible pipe. Keep in mind that it is easy to siphon out some fry also along with the waste. So, it is best to siphon out the waste into a clean bucket. This will help you to save any nosey fry.

## Failed Breedings

Reproductive failure is one of the most frustrating experiences for the aquarist, especially a beginner. These failures can occur at any stage of the reproductive cycle. The first major stumbling block for the beginner aquarist is failure to get the fish to spawn. For success, some species need the perfect set up.

A deep knowledge of the species, the right food, the right temperature for the species you select in addition to any necessary spawning triggers have to be used efficiently. For the beginner it is also important to choose the right species to breed as some species are much easier than others to breed. A separate spawning aquarium with or without plants according to the needs of your species is the first step in the right direction. Novice breeders also need to be very careful with the water chemistry. You cannot just put your selected pair in a tub of water and expect them to get on with their work. You have to get a fully working testing kit and learn how to use it. Higher quality in the breeding stock directly influences the viability of the spawn. The quality of the offspring will be affect by the kind of water in which it was bred.

Sexing the fish correctly is the next big step. In some species, differentiating between the two sexes is very difficult indeed. So, this is not to be taken lightly.

Do not keep checking on your fish every few minutes. This will make them stressed. Stressed fish will not spawn. Fish that are too young or too old must be avoided at all costs. Left to itself in nature, the female fish will select the male fish that is in its prime. Males that are too young or too old will not have viable sperms, and the eggs that are laid as a result may not get fertilized.

Once the fish has spawned, problems may arise at the time of hatching too. The biggest impediment to hatching is water quality. Bad water quality is not conducive to spawning, and even if the fish do spawn, the eggs will not develop properly.

As a result the eggs may catch fungus and die. If the eggs are removed to a hatching aquarium, care must be taken that they are not exposed to the atmosphere for long. The eggs dry up when they come into contact with air.

The water in the hatching aquarium must be as similar to that in the original aquarium as possible. Sudden changes in temperature will halt the development of the egg. Hard water will cause the shells of the eggs to harden, and in very hard water the shells can become too hard and make hatching difficult or impossible. If the water is too soft, the eggs may instead collapse.

Sterility is also a common cause for reproductive failure. In some species, it is seen that the absence of any male will cause two females to pair off. Obviously, no eggs will be fertile after such a spawning. Pairing off a female with a male that is too old or young will also cause the eggs to be underdeveloped. Again, hatching of such eggs will never take place.

Disease and vitamin deficiency are some other factors that play a decisive role in the ability of a fish to spawn. Over-feeding the fish and making them obese will rob some male species of their virility.

There are some varieties that are highly suited for the novice to breed. As stated earlier, livebearers are among the easiest to breed and will most likely spawn spontaneously in your aquarium without you even trying to coax them.

Of these, the Guppy, the Molly, the Platy and the Swordtail are the easiest to spawn. Mollies and Guppies are very similar in their breeding patterns. However, newborn Molly fry are susceptible to protozoan infection.

Sometimes, the Molly may also give birth to unformed fry that appear almost at the same time as the swimming fry. They appear as white eggs and must be disposed off without delay. Among egg-layers, the novice may want to try Steel Blue Killifish, some Dwarf Cichlids like Kribs, the Tiger Barbs, various Danios or the White Cloud Mountain Minnow.

Obviously, breeding fish is a reasonably challenging task, but it can be achieved quite easily, if the right steps are followed. Remember that as in humans the female carrying the eggs as well as the eggs themselves must be in the best possible surroundings. Maintaining the right condition inside the aquarium will help the dwellers of the aquarium to go about their work in a healthy manner.

## Feeding Fry

### *Infusorians*

Infusorians is a suitable food for very small fish fry. You can easily grow your own Infusoria cultures at home. This way you will always have cheap and nutritious food ready for your fry. Infusorians are minuscule single cell organisms that live in water. To cultivate your own Infusoria you will only need tap water, some bruised vegetables and a few clean containers such as jars or similar.

Boil the tap water and let it cool down to room temperature. Fill the jars with the water and add some bruised vegetables to each jar. Bruised lettuce leaves or banana skins are recommended, since they are known to produce a lot of Infusoria. One banana skin or 3-4 lettuce leaves is a good amount for each jar. Do not put any lids on the jars. Place your jars in a warm place where they will receive moderate light.

The water in the jars will soon turn cloudy and begin to smell a bit. As the Infusorians develop, the smell will become increasingly sweet and the water will clear up. It will usually take 3-4 days for Infusoria to develop. When the water looks clear, you can start feeding your fry with the Infusoria.

Use a siphon or a spoon to remove the Infusoria from the top of the jar and place it in your fry aquarium. Make sure not to bring any plant material from the Infusoria jars over to the fry rearing tank, since this can pollute the water. Start developing new Infusoria culture every 3-4 day to make sure you always have enough food for your fry until they are big enough to eat larger food.

### Brineshrimps (Artemia Salina)

Most fish fry is large enough to eat brine shrimp as their first food, especially if you give them newly hatched brine shrimp. Very small fry should instead be fed Infusoria, and brine shrimp can be introduced as a second food.

If you feed your brine shrimp suitable food, they will grow large enough to be used as food for adult fish as well. Growing large brine shrimps is an inexpensive way of providing your fish with much appreciated live food, and is especially recommended when you wish to induce spawning.

Just like Infusoria, brine shrimp is easy to cultivate at home. You can buy a brine shrimp hatchery from your fish store or fix one yourself. Brine shrimp will even hatch in bottles! Some brine shrimp hatcheries will require and air stone connected to an air operated pump. Other brine shrimp hatcheries can be attached to the inside or outside of the aquarium.

If you want to fix your own brine shrimp hatchery in a bottle you can use a clean milk bottle or similar. Boil some water, three-quarters of a pint is enough, and let it cool down before you use it to fill the bottle. Add one teaspoon of aquarium salt to the water and put the bottle in a warm room.

It is possible to use ordinary cooking salt instead of aquarium salt, but the result will usually not be as good as with aquarium salt. Insert an air stone in the bottle to provide oxygen. Place a pinch of brine shrimp eggs in the water and wait for them to hatch. The water temperature must never go below 15 degrees Celsius, but this is usually not a problem in a temperate room. Brine shrimp eggs will typically hatch after 48 hours or even earlier. By setting up new bottles you can make sure that you always have suitable size brine shrimp for your fry.

Brine shrimp eggs can be bought in most pet shops and fish stores. If you use eggs with shells you must turn off the air stone in the hatchery to make the shells sink to the bottom. The hatched brine shrimp will stay a few inches above the bottom and can be easily removed. It is important to separate the brine shrimp from the shells, since shells might get stuck inside the fish fry if they consume them. You can also buy shell-free brine shrimp eggs.

### *Water Fleas (Daphnia)*

Water fleas can be hard to disinfect properly, so if you want to feed your fry daphnia you must obtain the daphnia from a source that you can trust and that you know have no sanitary problems. The same is true if you plan to raise your own daphnia for the fry; you must make sure that you begin with a disease free starter culture.

Once your have obtained a proper starter culture of daphnia, the rest is easy. Fill a plastic container with 5 gallons of tap water and let the water stand for roughly 48 hours in room temperature. Obtain a handful of manure from a stable.

You will also need a nylon bag. You can produce one yourself from a pair of nylon stay-ups or use the type of nylon bag that comes with washing machine tablets. Place the manure in the nylon bag and submerge the nylon bag in the water. After roughly 7-10 days the water will have turned cloudy, which means that the water is now filled with micro organisms for the daphnia to feed on.

Add the daphnia to the water and wait. The culture will take care of itself for the few weeks needed for the daphnia to grow large and abundant enough to be suitable as fish food. The easiest way of removing daphnia is to use a fine meshed net.

### *Whiteworms*

Whiteworms are a much appreciated fry food that can be cultured at home. You will need to obtain some moist and nutritious garden soil for the whiteworms to grow in. Fill 75% of a shallow box with garden soil. If necessary, water the soil until it is quite damp.

It is important that the soil never is allowed to dry out, but waterlogged soil is just as bad. Add the whiteworm culture to the soil together with a few small pieces of moist white bread. Cover the box with a lid. The box or the lid must have some air holes to allow for ventilation. Put the box in a dark place where the temperature is around 16-18 degrees Celsius. Bring new bread to the box every 3-4 day and make sure that the soil is constantly moist. If you find any uneaten food you should remove it before it turns bad.

When its time for your to collect whiteworms for your fish, you just spoon out some soil from underneath the bread. You will naturally find the largest congregations of whiteworm right beneath the food pieces. Pour the soil with the whiteworms into a bowl or dish filled with water. When the worms have become separated from the soil you can easily collect them and drop them into the aquarium.

## Fish Spawning Methods

A majority of the world's fish species are egg layers, but there are also a lot of species that have developed into livebearers. The egg laying species will typically use some form of external fertilization of the eggs, while the livebearers perform internal fertilization. The level and type of parental care will also vary considerably between different fish species. Some are extremely devoted parents while others leave their offspring immediately, or even worse – try to eat them! The parental care can also vary a lot between different individuals within the same species, and from batch to batch.

### *Livebearers*

Livebearing fish species give birth to free swimming fry. A lot of the commonly kept aquarium species are livebearers, such as the Guppy. The eggs are fertilized by the male while they are still inside the body of the female fish, so called internal fertilization. In livebearing species, the male's anal fin has developed into a reproductive organ. This type of modified anal fin is called a gonopodium. After the fertilization, the eggs develop into fry inside the female fish. When the fry is born, they look like miniatures of adult fish, but sometimes without the striking colours of their parents. Many species develop striking colours as they grow larger and mature. In some livebearing species the female fish can store spermatozoa (semen) from one single mating and use it to fertilize several batches in a row. This is why female livebearers sometimes give birth in aquariums where they are kept without any males.

## *Egg Scatterers*

Egg scattering species are often living in fast flowing waters in the wild. The eggs are released by the female and fertilized, and then rapidly swept away by the current. This is why egg scattering species will usually eat their own offspring if they spawn in aquariums. If you are planning to breed an egg scatterer, e.g.a Zebra Danios, you must protect the eggs from the parents. You can cover the aquarium floor with marble and keep the water shallow. The eggs will sink and since the water is shallow they will rapidly reach the bottom, where they will lay hidden among the marbles. Larger marbles will also provide hiding places for the fry when they emerge. Some breeders place a net in the aquarium, to prevent the adult fish from reaching the bottom. The mesh must be large enough for the eggs to fall through.

## *Egg Depositers*

There are two types of egg depositers: open water spawners and secretive spawners. Open water spawners will not hide their eggs. Open water spawners typically deposit the eggs on a flat rock, on the aquarium floor or on the leaves of broad leaved aquatic plants. Some open water spawners will dig a crater in the substrate and place their eggs inside. Secretive egg depositers wants to hide their eggs, and will choose a cave, a crevice or similar as spawning site. In the aquarium they will usually appreciate turned over terracotta flowerpot. Both open water spawners and secretive spawners clean the site before they deposit their eggs, and they will usually also defend the eggs and care for them by cleaning them. Before the fry emerge, many egg depositers will move the eggs around to other pre-cleaned sites. Egg depositers are also known to guard their fry and the fry is often seen swimming around with their parents.

## *Nest Builders*

The nest building group is actually a sub-category to the egg depositers. Nest builders create nests of bubbles and deposit their eggs inside. The nests are formed by the male fish by blowing saliva bubbles, and a bubble nest will often contain plant fragments. When the female has released the eggs they are carefully gathered by the male and placed inside the bubble nest. In some species, the female will gather the eggs. In some species the eggs will float up into the nest without help from the parents.

### Mouth Brooders

Mouth brooders are also a kind of egg depositers, but the offspring is guarded inside the mouth of a parent. The eggs are often deposited in a shallow crater where they are fertilized by the male before being picked up by the male or the female fish. In some species the eggs are fertilized by the male when they have already been picked up by the female.

Some mouth brooding species will carry only the eggs inside the mouth and spit out the emerging fry, while others will care for the eggs on the spawning site and pick up the fry with their mouth when they emerge.

Some species will carry both egg and fry. Even after being released, the fry might flee into the mouth of their parent if they are scared by something or sleep inside the mouth during the night.

### Egg Buriers

The egg burying species have developed in areas subjected to seasonal draughts. During the rainy season, the eggs will be deposited deep down in the mud. The eggs will then stay buried during the draught. When a new rainy season begins, the water will trigger the hatching of the eggs and fry will emerge. The egg burying species will not guard their offspring, since the parents are usually dead long before the rainy season begins.

If you want to breed egg burying species in your aquarium you must simulate a dry season and a subsequent rainy season. The adult fish can be given a container filled with peat moss to deposit their eggs inside instead of placing them in mud. The container can be removed from the water and subjected to a several month long dry season in a cupboard. Once you add water to the peat moss again, the eggs will hatch and you can start feeding the fry.

## Koi Breeding

Koi breeding is a fascinating but can sometimes be quite hard since Koi fish grows quite large and because you will have to have good parents to produce quality Koi offspring.. This article is meant to explain some essential aspects in the breeding of Koi fish. All the described facts are very important so please don't mind the order in which they are written.

The first consideration when breeding Koi is to get suitable fish. High quality Koi will produce the highest quality offspring and you will have to choose the parents after which kind of offspring you want in regards to colouration and pattern. Only a very low percentage of the fry in the large batches will become high quality Koi regardless of the quality of the parents.

High quality parents will however produce a higher percentage quality fry. It is recommendable to try and keep the bloodlines of you Koi type clean to produce high quality Koi fish.

Another aspect that should be taken into consideration while choosing the indented parents is the age of the fish. As this effects the quality of the egg shells. Young Koi females produce eggs with very thin shells which might not survive; older fish on the other hand can produce eggs with to hard shells that the sperm can't penetrate them. The best age of a Koi that is used for breeding is 4-5 years although successful breeding might be possible with fish that are up to 15 years old.

The chosen couple should be separated from other fish of the other sex for at least a month during which time they are kept well feed to get them into spawning condition. Koi breeding is often best done with pairs of fish but you can use 2 males to a female while breeding Koi. You should however never use more than two males since they may become too aggressive and hurt the female if more than two males are used.

The best time of the year to breed Koi if it is done in out doors is different depending on where you live in the world it is however recommendable to use a time of the year when the temperature difference between different times of the day is as smallest. Koi breeding are most likely to be successfully during a full moon since this helps trigger the fish to spawn.

### Koi Breeding Tank

You will have to decorate the aquarium so that there are suitable breeding areas present for the Koi fish. The materials selected to be used for spawning should be totally submerged and safe to the environment. There are many synthetic Koi breeding materials available in the trade but most professional breeders still prefer the natural ones and think the produce better results. An example of a good natural breeding material is Water hyacinths which are a plant that can be great for building suitable spawning places with. Water hyacints can carry parasites and will have to treated with special substances like Dipterex to eliminate all possible parasites before being used in a Koi breeding setup. Water hyacinths might also attract dragon flies. While these insects are harmless, their larvae can attack and kill the Koi eggs and fry. It is recommendable to use nettings on top of your breeding pond if you breed your Koi outdoors to protect them from predators. It might also be beneficial since they sometimes like to jump and netting prevents them from jumping out of the pond. A benefit that makes the use of nettings on your pond a beneficial choice regardless of whether you keep and breed your Koi indoors or outdoors.

The pH level in the tank is not that important in regards to getting the Koi to spawn but will effect the quality of the fry. A neutral water is to be preferred as acidic water causes the fish to get good red colour but weak black colour and an alkaline water will cause good black colour put weak red colour.

The good Koi breeding pond should have a round shape and around 1,000 gallons / 3800 L of fresh and chlore-free water. The water will if Chlorated have to be treated with Clorox or another equivalent to remove the chlore before being used to breed koi. The water should be kept standing for three days before adding fish. The water should be aerated using some strong water pumps. A strong current or a splashing waterfall adds to the aeration but can have a damaging effect on the spawning.

The male will try to push the female against the pond walls and other objects to force the eggs out of here during the spawning. He fertilizes the eggs once they are released. If two males are used they often push the female from each side to get her to release her eggs. One spawning results in many thousand sticky eggs that adhere to anything they touch. It is good to keep a watchful eye on the spawning since the male sometimes injures the female in which case you will need rescuing. The parents will eat their own egg and fry so the It is recommendable to move the male and the female to different tanks and to give the female a chance to rest after the breeding. The female will still smell like spawning and males will keep trying to spawn with her if you are kept in tanks containing males. This can often lead to the males hurting the female. The spawning smell will have vanished after about 24 hours. The male will also have to recuperate since he otherwise might become infected with a diseases and fungus

The fry will survive on their yolk sac their first day but will after that need to be feed 4-5 times a day for the following 3 months. Suitable food includes infusoria and algae (supplement food only) and later newly hatched brine shrimp.

As the fry grow you can start switching them over to regular Koi pellets. Make sure that you keep the water in good condition as the frequent feedings put a large strain on the water.

It is often almost impossible to raise all the fry in a spawning and it is recommendable to only raise half the fry or to move half of them to another pond since the breeding pond as described above otherwise becomes too crowded. It is also recommend to sort the fish after quality as they grow and only keep the fry of higher quality.

## Raising Fry

Once the eggs are laid, you can remove the parents in none fry guarding species. This is especially important if the fish are the types that eat their young. In case of fish that care for their brood, it is recommended to keep the parents with the fry.

Taking care of your fry requires some special attention. Some young ones, like the Goldfish, will not eat for the first 40-48 hours of their life, while other species need to be fed immediately. How do you know when to do what? And what are the best conditions for keeping the fry alive. These are some of the many things that you need to keep in mind when the fish start spawning.

In some livebearers, the female may release only a few of the babies at the beginning. She may take a few hours or even a day to release the entire batch of fry; and some of these fry will likely be stillborn or even eggs.

Even after this, a female can be full of fertilized eggs that will soon begin to develop into fish. Livebearers can release several more batches of fry even if there are no males around, due to their ability to save sperm.

The fry may take up to a few weeks to develop. Do not expect your fry to come out ready for a swim and a feed. The larvae that hatch have almost nothing in common with their parents. They will be enclosed in a big yellow yolk sac and will not be able to swim at all. The larvae will eat up the egg sac until the yolk is fully gone. Once this is over, the fry begin to get hungry and look for food.

Infusoria is a good food for tiny fry. Green water or egg yolk can also be fed to the fry. Powdered flake food is commonly fed to fish fry in aquariums, but powdered flake food is not the best fry food. Powdered flake food is however better than nothing if it is all you got when the fry arrives. Take a plastic bag and put in 1-2 tablespoons of flake food. The bag should not have too much air in it. Then squeeze the bag in your hand and mash the flake as well as you can. The more powdered it is, the better.

Putting in only a small amount of food for the fry is also a bit difficult. The best way to add in small amounts of powdered food is by using a toothpick. Wet the end of the pick in water, and wipe it with your finger so that it is only a little wet. The wet toothpick should now be dipped into the powdered flakes. Touch this tip in the water just above the fry. You will need to feed them several times a day.

This powder is quite fine and is acceptable for many fry to start off with. But some fry find even this too big to eat. For instance, the newly hatched Angelfish cannot eat this powder. It needs to eat infusoria or baby brine shrimp for the first few days.

Baby brine shrimp is a very important food for most fish fry. Most livebearers will eat lots of baby brine shrimp, and this definitely makes them grow faster and healthier even if livebearers are among the fry that can be raised on powdered flake food alone. The eggs of Brine Shrimp are carefully processed and collected and are sold in many pet stores. The aquarist can purchase these eggs, hatch them and use them to feed fry. One advantage in hatching the eggs at home is that these can then be fed special additives that will directly be transferred to your fry. Some fry are too small to be able to eat newly hatched brine shrimp and such fry will need infusoria the first few days before they can start eating brine shrimp.

After they start feeding, the fry usually grow pretty fast. Feeding should be done several times a day for optimal growth. Aquariums that are well lit, well planted and nicely filtered provide the best conditions for rearing fry. Having a fixed feeding schedule also helps the fry grow faster. Frequent water changes are a MUST if you want healthy fry. Remember that the filtration in an aquarium that contains small fry will be less efficient than the filtration in a proper aquarium as the circulation has to be kept lower. The number of feedings is also higher and the wastage will also be more. Thus frequent water changes; stable water chemistry and sufficient warmth contribute greatly towards the well being of your fry.

## Spawning Methods

Sexually, fish can be classified into two groups: the sexually dimorphic and the sexually isomorphic species. The sexually dimorphic species show obvious primary and secondary characteristics of their gender. The males of this species are often larger and more colourful, and they usually have a much more elaborate finage.

The sexually isomorphic species show hardly any apparent differences in the two sexes. It is therefore very difficult to distinguish between the two sexes in this group. Often, the only deciding factor could be the shape of the genital papilla. This becomes visible at the time of spawning. In some of these species, the female have a more rounded belly or is slightly larger than the males. There are some species that show no visible difference at all between the two sexes.

External fertilization followed by egg laying is the way most fish in aquariums will reproduce. Egg layers are classified into five groups:

### *Egg Scatterers*

These species, as the name indicates, scatter their eggs on any surface. The eggs may be adhesive or non adhesive. After laying eggs, the parents do not look after them, and may even eat their own eggs. These are mostly schooling fish, and they often spawn in groups. Their eggs hatch fairly quickly.

### *Egg Depositors*

These fish deposit their eggs somewhere on the substrate. They lay fewer eggs than egg scatterers. Some species within this group show advanced care for the eggs and fry. They will deposit eggs in an enclosure, cave, a pit or similar, and protect and clean the eggs. Some species of Cichlids come under this category. Not all egg depositing species care for their young. Some just deposit their eggs and then abandon them. Egg depositing species generally don't eat their own eggs. Their eggs are usually larger than those of the scatterers.

### Mouth Brooders

There are two basic types of mouth brooders. In some species, the female fish carries the eggs and/or larvae in its mouth. The female in this group receives a mouthful of sperm from the male, and the fertilization takes place in the mouth of the female. In the second type of mouthbreeders, the females lay the eggs on a substrate and guard these eggs till they hatch. Once the eggs hatch, she picks up the fry and carries them in her moth till the fry have become a little bigger. There are also paternal mouthbrooders where the offspring is carried by the male fish.

### Nest Builders

These fish build a kind of nest that is made of plant debris and saliva. They may also dig a pit in the substrate to lay eggs. They take care of their young ones.

### Livebearers

Livebearers are fish that do not lay eggs. The fry come already hatched out of the mother fish. Livebearers are able to give birth to young ones because of internal fertilization which takes place inside the female. Sometimes contact between the cloacal apertures of the male and female fish leads to fertilization. In some species, the eggs produced by the livebearer will be heavily yoked. In other species, the young are in direct contact with the mother and are fed and nourished by her. In this case, the mother carrying the young ones needs more nutrition and overall care. Fry of livebearing fish usually takes about 30-32 days to develop. Many factors will affect this development period, e.g. the water temperature, the female's age and her level of nutrition. The good thing about livebearer females is that they are immediately ready to breed again after they drop their young ones. The competition to get to these females is intense among the males.

For beginners livebearers are the easiest fish to breed. If given enough space, the livebearers will do most of the hard work. Once the fry become noticeable and have started swimming, they need to be removed from the aquarium unless the aquarium is densely planted. Failure to do so will surely land a large percentage of them into the stomachs of bigger fish. Many people breed livebearers to provide food for predatory fish. Some livebearers will cross breed with other fish and to avoid hybridisation these species should not be kept together.

## Tips for Effective Fish Breeding

Anything that you have placed in the aquarium traps debris. As soon as the fry come out, remove any breeding traps that have been in the aquarium before. You can also add one or a couple of apple snails into the aquarium. Apple snails are very laid back and do not attack any fish. These snails will eat debris, and will also eat any dead fry. Apple snails will not eat live fry. Apple snails also produce a good supply of infusorians and can even provide the frys first supply of food.

Keeping a detailed log of the entire process is a very good idea. This will help you when you need to repeat the process. You should keep accounts of the species name, the detailed water chemistry, filtration and aeration methods, approximate ages of the parents, when the female was added, the date on which the fry was released/ hatched, the frys first food and the body size of the fry till it reaches about three months of age. Sometimes, even in spite of doing just about everything to ensure success, you will find that your fish just does not breed. All conditions are as they should be, but still there is no success. In this case, it is best to use a target fish to jump-start the process. This works best in territorial fish. A target fish is a perceived threat to the territorial fish. The male fish sees the new fish as a possible encroacher in his territory and he will then pair up with the female fish to isolate the target fish.

The important thing here is to ensure that your target fish is in reality not a threat to the existing fish or vice versa. For instance, when trying to breed small cichlids, using some zebra danios as target fish will not pose any threat to either fish.

It should be noted that some species of fish will kill any kind of target fish, and such situations should be avoided. If this is the case, it will also suffice if you place the target fish in a different aquarium alongside the mating aquarium so that the male can see but not touch.

Sometimes, fish introduced into a new aquarium are too nervous to come out into the open. They will remain hidden for days and will refuse to settle down. Fish that are stressed in this way will never pair up and breed. In such cases, you can use a dither fish to calm the other fish.

A dither fish is an easygoing, harmless fish. A hyperactive and aggressive fish is definitely not the right dither fish. Calm and peaceful midwater fish makes the best dither fish. Once the nervous fish see the dither fish swimming about happily without being consumed by predators, they too will settle down and come out of their hiding places. This is just to reassure the nervous fish that nothing will harm them when they come out into the open.

## Breeding Techniques for Tropical Fish

Breeding aquarium fish is one of the steps to becoming a skilled aquariast. Once fish are kept healthy enough to spawn, the novice knows that they can attempt to keep more challenging fish. In order to breed a species, the aquariast usually needs to be able to distinguish between the sexes and to be able to recreate natural conditions to stimulate spawning. Always record your successes and failings in a notebook.

### *Sexing Fish*

Determining the sex of a fish is an important step in knowing whether one has a pair. Most fish can be classified as sexually dimorphic or sexually isomorphic. In sexually dimorphic species, the sexes can be easily distinguished by primary (shape of sex organs) and secondary differences (size, shape, colour [sexual dichromatism], finnage). Males are frequently more colourful, larger, and have more elaborate finnage. Among the more brilliant outstanding of sexual dimorphism can be found in Lake Malawi cichlids, Killifish, and Livebearers. In sexually isomorphic species, there are minute, if any, apparent sexual differences. Often, the only way to distinguish between the sexes is the shape of the genital papilla, which is only visible around spawning times. In some isomorphic species, the males are slightly larger and the females are slightly rounder in the belly. Some sexually isomorphic species have no known external sexual differences.

### *Selecting the Parent Fish*

Once males and females have been distinguished, a suitable pair or spawning group should be chosen. There are several important traits to seek in choosing the parent fish.

1. Choose fish that display good markings and colour, that should produce attractive young.
2. Only use mature, healthy fish for spawning because unhealthy fish, if they will spawn, may produce unhealthy or deformed young.

3. Be sure that the pair is compatible. Many species cannot be put together in a breeding tank and expected to get along and produce young. In fact with many cichlids, pairs form only after a group has been raised together for months if not years. In certain species, one partner will bully the other to death if there is not compatibility.
4. Avoid crossing different strains or colour forms because the young are often unattractive.
5. Make sure that the pair are both of the same species because hybrids are sterile. With some cichlids and Killifish, females of different species look similar.

### Reproductive Strategies Egg-layers

The majority of aquarium fish are egg-layers with external fertilization. Egg-layers can be divided into five groups: egg-scatterers, egg-depositors, egg-burriers, mouth-brooders, and nest-builders.

*Egg-scatterers:* These species simply scatter their adhesive or non-adhesive eggs to fall to the substrate, into plants, or float to the surface. These species do not look after their brood and even eat their own eggs. These, often schooling, fish may spawn in groups or in pairs. Often there is a large number of the small eggs laid. The fry hatch quickly.

*Egg-depositors:* These species deposit their eggs on a substrate (tank glass, wood, rocks, plants). Egg depositors usually lay less eggs than egg-scatterers, although the eggs are larger.

*Egg-depositors fall into two groups:* those that care for their eggs, and those that do not. Among eggs depositors that care for their eggs are cichlids and some catfish.

*Egg-depositors that care for their young can be divided into two groups:* cavity spawners and open spawners. Cavity spawners lay their eggs in a cave, while open (shelter) spawners lay their eggs on an open surface. These fish form pairs and have advanced brood care where the eggs are defended and cleaned. The eggs take a few days to hatch, and the fry are often guarded by the parents. Various catfish, Cyprinds, and Killifish make up the majority of egg-depositors that do not care for their young. These species lays their eggs against a surface, where the eggs are abandoned. These species do not usually eat their eggs.

*Egg-burriers:* These species usually inhabit waters that dry up at some time of the year. The majority of egg burriers are annual Killifish which lay their eggs in mud. The parents mature very quickly and lay their eggs before dying when the water dries up. The eggs remain in a dormant stage until rains stimulate hatching.

*Mouth-brooders:* Are species that carry their eggs or larvae in their mouth. Mouth brooders can be broken up into ovophiles and larvophiles. Ovophile or egg-loving mouth-brooders lay their eggs in a pit, which are sucked up into the mouth of the female. The small number of large eggs hatch in the mother's mouth, and the fry remain there for a period of time. Fertilization often occurs with the help of egg-spots, which are colourful spots on the anal fin of the male. When the female sees these spots, she tries to pick up the egg-spots, but instead gets a mouthful of sperm, fertilizing the eggs in her mouth. Many cichlids and some labyrinth fish are ovophile mouthbrooders. Larvophile or larvae-loving mouth-brooders lay their eggs on a substrate and guard them until the eggs hatch. After hatching, the female picks up the fry and keeps them in her mouth. When the fry can fend for themselves, they are released. Some eartheaters are larvophile mouthbrooders.

*Nest-builders:* Nest builders build some sort of nest for their eggs. The nest is usually in the form of bubble-nest formed with plant debris and saliva-coated bubbles (labyrinth fish, catfish), or a excavated pit in the substrate (cichlids). Nest builders practice brood care.

# 2

# Characteristics of Fish Farming

## Channel Catfish

Channel catfish, *Ictalurus punctatus*, is North America's most numerous catfish species. It is the official fish of Missouri, Iowa, Nebraska, Kansas, and Tennessee, and is informally referred to as a "channel cat". In the United States they are the most fished catfish species with approximately 8 million anglers targeting them per year. The popularity of channel catfish for food has contributed to the rapid growth of aquaculture of this species in the United States.

### *Distribution and Habitat*

Channel catfish are native to the Nearctic, being well distributed in lower Canada and the eastern and northern United States, as well as parts of northern Mexico. They have also been introduced into some waters of landlocked Europe and parts of Malaysia and almost many parts of Indonesia. They thrive in small and large rivers, reservoirs, natural lakes, and ponds. Channel "cats" are cavity nesters, meaning they lay their eggs in crevices, hollows, or debris, in order to protect them from swift currents. In Canada, the species is largely, though not exclusively, limited to the Great Lakes watershed from Lake Nipigon southward.

### *Characteristics*

Channel catfish possess very keen senses of smell and taste. At the pits of their nostrils (nares) are very sensitive odour sensing organs with a very high concentration of olfactory receptors. In channel catfish these organs are sensitive enough to detect several amino acids at about 1 part per 100 million in water. In addition, channel catfish have taste buds distributed over the surface of their entire body. These buds are especially concentrated on the channel catfish's four pair of barbels (whiskers) surrounding the mouth — about 25 buds per square millimetre. This combination of exceptional senses of taste and smell allows the channel catfish to find food in dark, stained, or muddy water with relative ease.

***Figure:*** *Chuck the Channel catfish, 1986 roadside sculpture in Selkirk, ManitobaCanada*

### Fishing

Channel catfish are omnivores and can be caught using a variety of natural and prepared baits, including crickets, night crawlers, minnows, shad, crawfish, frogs, bullheads, sunfish, and suckers. Catfish have even been known to take Ivory Soap as bait. Another method of catching catfish is using stinkbaits, which are prepared with dead fish, crawfish, garlic, blood, liver, meat, cheese, dough, and even Kool-Aid powder. Sometimes these stinkbaits are prepared into a doughball and mashed onto a hook; other times they are smeared in special tubes meant to hold these baits, and fished slowly on the bottom. Grocery store baits such as chicken livers, shrimp, dog food, squid, and bubble gum will also catch plenty of channel cats.

Juglines, trotlines, limb lines and bank lines are popular methods of fishing for channel catfish in addition to traditional rod and reel fishing. Another method uses traps, either "slat traps" — long wooden traps with an angled entrance — and wire hoop traps. Typical bait for these traps include rotten cheese and dog food. Catches of as many as 100+ fish a day are common in catfish traps.

When removing the hook from a catfish, anglers should be mindful of the sharp spines on the pectoral and dorsal fin.

### Length and Weight

A member of the *Ictalurus* genus of American catfishes, channel catfish have a top-end size of approximately 40-50 pounds (18–23 kg). The world record channel catfish weighed 58 pounds and was taken from the Santee-Cooper Reservoir in South Carolina, July 7, 1964. Realistically, a channel catfish over 20 pounds (9 kg) is a spectacular specimen, and most catfish anglers view a 10 pound (4.5 kg) fish as a very admirable catch. Furthermore the average size channel catfish an angler could expect to find in most waterways would be between 2 and 4 pounds.

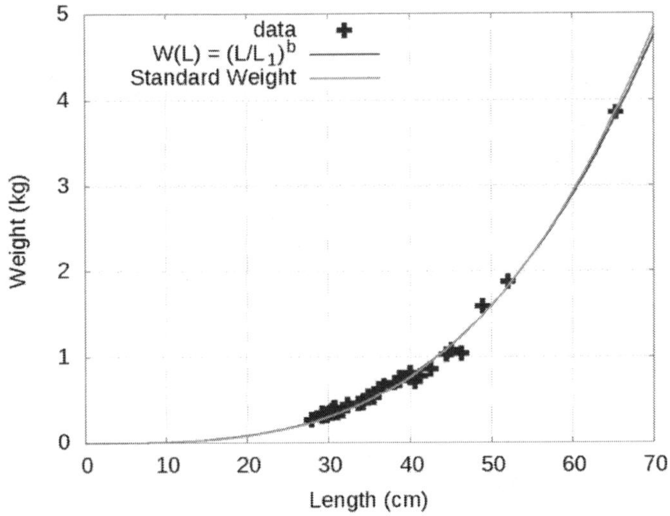


***Figure:*** *Weight vs. length for Channel Catfish, where b = 3.2293 and* $L_1 = 45.23$ *cm.*

Channel catfish will often coexist in the same waterways with its close relatives, blue catfish, which are somewhat less common but tend to grow a lot larger (with several specimen confirmed to weight above the 100 lb. mark). As channel catfish grow longer, they increase in weight. The relationship between length and weight is not linear, but instead follows a pareto law.

The relationship between length (L, in cm) and weight (W, in kg) for nearly all species of fish can be expressed by an equation of the form:

$$W = (L / L_1)^b$$

Invariably, b is close to 3.0 for all species, $L_1$ is the length of a typical fish weighing 1 kg. For channel catfish, b = 3.2293, somewhat higher than for many common species, and $L_1 = 45.23$ cm.

## Aquaculture

Aquaculture, also known as aquafarming, is the farming of aquatic organisms such as fish, crustaceans, molluscs and aquatic plants. Aquaculture involves cultivating freshwater and saltwater populations under controlled conditions, and can be contrasted with commercial fishing, which is the harvesting of wild fish. Mariculture refers to aquaculture practiced in marine environments.

The reported output from global aquaculture operations would supply one half of the fish and shellfish that is directly consumed by humans; however, there are issues about the reliability of the reported figures. Further, in current aquaculture practice, products from several pounds of wild fish are used to produce one pound of a piscivorous fish like salmon.

Particular kinds of aquaculture include fish farming, shrimp farming, oyster farming, algaculture (such as seaweed farming), and the cultivation of ornamental fish. Particular methods include aquaponics and Integrated multi-trophic aquaculture, both of which integrate fish farming and plant farming.

### *History*

The indigenous Gunditjmara people in Victoria, Australia may have raised eels as early as 6000 BC. There is evidence that they developed about 100 square kilometres (39 sq mi) of volcanic floodplains in the vicinity of Lake Condah into a complex of channels and dams, that they used woven traps to capture eels, and that capturing and smoking eels supported them year round.

Aquaculture was operating in China circa 2500 BC. When the waters subsided after river floods, some fishes, mainlycarp, were trapped in lakes. Early aquaculturists fed their brood using nymphs and silkworm feces, and ate them. A fortunate genetic mutation of carp led to the emergence of goldfish during the Tang Dynasty. Japanese cultivated seaweed by providing bamboo poles and, later, nets and oyster shells to serve as anchoring surfaces for spores.

***Figure:*** *Workers harvest catfish from the Delta Pride Catfish farms in Mississippi*

***Romans Bred Fish in Ponds:*** In central Europe, early Christian monasteries adopted Roman aquacultural practices. Aquaculture spread in Europe during the Middle Ages, since away from the seacoasts and the big rivers, fish were scarce/expensive. Improvements in transportation during the 19th century made fish easily available and inexpensive, even in inland areas, making aquaculture less popular.

Hawaiians constructed oceanic fish ponds. A remarkable example is a fish pond dating from at least 1,000 years ago, at Alekoko. Legend says that it was constructed by the mythical Menehune dwarf people.

In 1859 Stephen Ainsworth of West Bloomfield, New York, began experiments with brook trout. By 1864 Seth Green had established a commercial fish hatching operation at Caledonia Springs, near Rochester, New York.

By 1866, with the involvement of Dr. W. W. Fletcher of Concord, Massachusetts, artificial fish hatcheries were under way in both Canada and the United States.

When the Dildo Island fish hatchery opened in Newfoundland in 1889, it was the largest and most advanced in the world. Californians harvested wild kelp and attempted to manage supply circa 1900, later labelling it a wartime resource.

### Definition

***Figure:*** *Tilapia, a commonly farmed fish due to its adaptability*

According to the FAO, aquaculture "is understood to mean the farming of aquatic organisms including fish, molluscs, crustaceans and aquatic plants. Farming implies some form of intervention in the rearing process to enhance production, such as regular stocking, feeding, protection from predators, etc. Farming also implies individual or corporate ownership of the stock being cultivated."

### 1st Century Practice

About 430 (97%) of the species cultured as of 2007 were domesticated during the 20th century, of which an estimated 106 came in the decade to 2007. Given the long-term importance of agriculture, it is interesting to note that to date only 0.08% of known land plant species and 0.0002% of known land animal species have been domesticated, compared with 0.17% of known marine plant species and 0.13% of known marine animal species. Domestication typically involves about a decade of scientific research. Domesticating aquatic species involves fewer risks to humans than land animals, which took a large toll in human lives. Most major human diseases originated in domesticated animals, through diseases such as smallpox and diphtheria, that like most infectious diseases, move to humans from animals. No human pathogens of comparable virulence have yet emerged from marine species.

Harvest stagnation in wild fisheries and overexploitation of popular marine species, combined with a growing demand for high quality protein, encourage aquaculturists to domesticate other marine species.

### Production Volume

***Table:*** *Top ten species groups in 2004*

| *Species Group* | *Million Tonnes* |
|---|---|
| Carps and other cyprinids | 18.30 |
| Oysters | 4.60 |
| Clams, cockles, ark shells | 4.12 |
| Miscellaneous freshwater fishes | 3.74 |
| Shrimps, prawns | 2.48 |
| Salmons, trouts, smelts | 1.98 |
| Mussels | 1.86 |
| Tilapias and other cichlids | 1.82 |
| Scallops, pectens | 1.17 |
| Miscellaneous marine molluscs | 1.07 |

In 2004, the total world production of fisheries was 140 million tonnes of which aquaculture contributed 45 million tonnes, about one third. The growth rate of worldwide aquaculture has been sustained and rapid, averaging about 8 percent per annum for over thirty years, while the take from wild fisheries has been essentially flat for the last decade. The aquaculture market reached $86 billion in Aquaculture is an especially important economic activity in China. Between 1980 and 1997, the Chinese Bureau of Fisheries reports, aquaculture harvests grew at an annual rate of 16.7 percent, jumping from 1.9 million tonnes to nearly 23 million tonnes. In 2005, China accounted for 70% of world production. Aquaculture is also currently one of the fastest growing areas of food production in the U.S.

***Table:*** *Top ten aquaculture producers in 2004*

| ***Country*** | ***Million Tonnes*** |
|---|---|
| China | 30.61 |
| India | 2.47 |
| Vietnam | 1.20 |
| Thailand | 1.17 |
| Indonesia | 1.05 |
| Bangladesh | 0.91 |
| Japan | 0.78 |
| Chile | 0.67 |
| Norway | 0.64 |
| United States | 0.61 |
| Other countries | 5.35 |
| Total | 45.47 |

Approximately 90% of all U.S. shrimp consumption is farmed and imported. In recent years salmon aquaculture has become a major export in southern Chile, especially in Puerto Montt, Chile's fastest-growing city.

### Over Reporting

China overwhelmingly dominates the world in reported aquaculture output. They report a total output which is double that of the rest of the world put together. However, there are issues with the accuracy of China's returns.

In 2001, the fisheries scientists Reg Watson and Daniel Pauly expressed concerns in a letter to *Nature*, that China was over reporting its catch from wild fisheries in the 1990s. They said that made it appear that the global catch since 1988 was increasing annually by

300,000 tonnes, whereas it was really shrinking annually by 350,000 tonnes. Watson and Pauly suggested this may be related to China policies where state entities that monitor the economy are also tasked with increasing output. Also, until recently, the promotion of Chinese officials was based on production increases from their own areas.

China disputes this claim. The official Xinhua News Agency quoted Yang Jian, director general of the Agriculture Ministry's Bureau of Fisheries, as saying that China's figures were "basically correct". However, the FAO accepts there are issues with the reliability of China's statistical returns, and currently treats data from China, including the aquaculture data, apart from the rest of the world.

## Methods

### *Mariculture*

Mariculture is the term used for the cultivation of marine organisms in seawater, usually in sheltered coastal waters. In particular, the farming of marine fish is an example of mariculture, and so also is the farming of marine crustaceans (such as shrimps), molluscs (such as oysters) and seaweed.

### *Integrated*

Integrated Multi-Trophic Aquaculture (IMTA) is a practice in which the by-products (wastes) from one species are recycled to become inputs (fertilizers, food) for another. Fed aquaculture (for example, fish, shrimp) is combined with inorganic extractive (for example, seaweed) and organic extractive (for example, shellfish) aquaculture to create balanced systems for environmental sustainability (biomitigation), economic stability (product diversification and risk reduction) and social acceptability (better management practices).

"Multi-Trophic" refers to the incorporation of species from different trophic or nutritional levels in the same system. This is one potential distinction from the age-old practice of aquatic polyculture, which could simply be the co-culture of different fish species from the same trophic level.

In this case, these organisms may all share the same biological and chemical processes, with few synergistic benefits, which could potentially lead to significant shifts in the ecosystem. Some traditional polyculture systems may, in fact, incorporate a greater diversity of species, occupying several niches, as extensive cultures (low intensity, low management) within the same pond. The "Integrated" in IMTA refers to the more intensive cultivation of the different species in proximity of each other, connected by nutrient and energy transfer through water.

Ideally, the biological and chemical processes in an IMTA system should balance. This is achieved through the appropriate selection and proportions of different species providing different ecosystem functions. The co-cultured species are typically more than just biofilters; they are harvestable crops of commercial value.

A working IMTA system can result in greater total production based on mutual benefits to the co-cultured species and improved ecosystem health, even if the production of individual species is lower than in a monoculture over a short term period.

Sometimes the term "Integrated Aquaculture" is used to describe the integration of monocultures through water transfer. For all intents and purposes however, the terms "IMTA" and "integrated aquaculture" differ only in their degree of descriptiveness.

Aquaponics, fractionated aquaculture, IAAS (integrated agriculture-aquaculture systems), IPUAS (integrated peri-urban-aquaculture systems), and IFAS (integrated fisheries-aquaculture systems) are other variations of the IMTA concept.

### Netting Materials

Various materials, including nylon, polyester, polypropylene, polyethylene, plastic-coated welded wire, rubber, patented rope products (Spectra, Thorn-D, Dyneema), galvanised steel and copper are used for netting in aquaculture fish enclosures around the world. All of these materials are selected for a variety of reasons, including design feasibility, material strength, cost, and corrosion resistance.

***Copper Alloys in Aquaculture:*** Recently, copper alloys have become important netting materials in aquaculture because they are antimicrobial (i.e., they destroy bacteria, viruses, fungi, algae, and other microbes) and they therefore prevent biofouling (i.e., the undesirable accumulation, adhesion, and growth of microorganisms, plants, algae, tubeworms, barnacles, mollusks, and other organisms). By inhibiting microbial growth, copper alloy aquaculture cages avoid costly net changes that are necessary with other materials. The resistance of organism growth on copper alloy nets also provides a cleaner and healthier environment for farmed fish to grow and thrive.

## Species Groups

### *Fish*

The farming of fish is the most common form of aquaculture. It involves raising fish commercially in tanks, ponds, or ocean enclosures, usually for food. A facility that releases juvenile fish into the wild for recreational fishing or to supplement a species' natural numbers is generally referred to as a fish hatchery. Fish species raised by fish farms include salmon, bigeye tuna, carp, tilapia, catfish and cod.

In the Merranean, young bluefin tuna are netted at sea and towed slowly towards the shore. They are then interned in offshore pens where they are further grown for the market. In 2009, researchers in Australia managed for the first time to coax tuna (Southern bluefin) to breed in landlocked tanks.

### *Crustaceans*

Commercial shrimp farming began in the 1970s, and production grew steeply thereafter. Global production reached more than 1.6 million tonnes in 2003, worth about 9 billion U.S. dollars. About 75% of farmed shrimp is produced in Asia, in particular in China and Thailand. The other 25% is produced mainly in Latin America, where Brazil is the largest producer. Thailand is the largest exporter.

Shrimp farming has changed from its traditional, small-scale form in Southeast Asia into a global industry. Technological advances have led to ever higher densities per unit area, and broodstock is shipped worldwide. Virtually all farmed shrimp are penaeids (i.e., shrimp of the family *Penaeidae*), and just two species of shrimp, the Pacific white shrimp and the giant tiger prawn, account for about 80% of all farmed shrimp. These industrial monocultures are very susceptible to disease, which has decimated shrimp populations across entire regions. Increasing ecological problems, repeated disease outbreaks, and pressure and criticism from both NGOs and consumer countries led to changes in the industry in the late 1990s and generally stronger regulations. In 1999, governments, industry representatives, and environmental organisations initiated a program aimed at developing and promoting more sustainable farming practices.

Freshwater prawn farming shares many characteristics with, including many problems with marine shrimp farming. Unique problems are introduced by the developmental life cycle of the main species, the giant river prawn.

The global annual production of freshwater prawns (excluding crayfish and crabs) in 2003 was about 280,000 tonnes of which China produced 180,000 tonnes followed by India and Thailand with 35,000 tonnes each. Additionally, China produced about 370,000 tonnes of Chinese river crab.

### Molluscs

Abalone farming began in the late 1950s and early 1960s in Japan and China. Since the mid-1990s, this industry has become increasingly successful. Over-fishing and poaching have reduced wild populations to the extent that farmed abalone now supplies most abalone meat.

### Echinoderms

Commercially harvested echinoderms include sea cucumbers and sea urchins. In China, sea cucumbers are farmed in artificial ponds as large as 1,000 acres (400 ha).

### Algae

Microalgae, also referred to as phytoplankton, microphytes, or planktonic algae constitute the majority of cultivated algae. Macroalgae, commonly known as seaweed, also have many commercial and industrial uses, but due to their size and specific requirements, they are not easily cultivated on a large scale and are most often taken in the wild.

### Issues

Aquaculture can be more environmentally damaging than exploiting wild fisheries on a local area basis but has considerably less impact on the global environment on a per kg of production basis. Local concerns include waste handling, side-effects of antibiotics, competition between farmed and wild animals, and using other fish to feed more marketable carnivorous fish. However, research and commercial feed improvements during the 1990s & 2000s have lessened many of these.

Fish waste is organic and composed of nutrients necessary in all components of aquatic food webs. In-ocean aquaculture often produces much higher than normal fish waste concentrations. The waste collects on the ocean bottom, damaging or eliminating bottom-dwelling life. Waste can also decrease dissolved oxygen levels in the water column, putting further pressure on wild animals.

### Fish Oils

Tilapia from aquaculture has been shown to contain more fat and a much higher ratio of omega-6 to omega-3 oils.

### Impacts on Wild Fish

Salmon farming currently leads to a high demand for wild forage fish. Fish do not actually produce omega-3 fatty acids, but instead accumulate them from either consuming microalgae that produce these fatty acids, as is the case with forage fish like herring and sardines, or, as is the case with fatty predatory fish, like salmon, by eating prey fish that have accumulated omega-3 fatty acids from microalgae. To satisfy this requirement, more than 50 percent of the world fish oil production is fed to farmed salmon.

In addition, as carnivores, salmon require large nutritional intakes of protein, protein which is often supplied to them in the form of forage fish. Consequently, farmed salmon consume more wild fish than they generate as a final product. To produce one pound of farmed salmon, products from several pounds of wild fish are fed to them.

As the salmon farming industry expands, it requires more wild forage fish for feed, at a time when seventy five percent of the worlds monitored fisheries are already near to or have exceeded their maximum sustainable yield. The industrial scale extraction of wild forage fish for salmon farming then impacts the survivability of the wild predator fish who rely on them for food.

Fish can escape from coastal pens, where they can interbreed with their wild counterparts, diluting wild genetic stocks. Escaped fish can become invasive, out competing native species.

### Coastal Ecosystems

Aquaculture is becoming a significant threat to coastal ecosystems. About 20 percent of mangrove forests have been destroyed since 1980, partly due to shrimp farming. An extended cost–benefit analysis of the total economic value of shrimp aquaculture built on mangrove ecosystems found that the external costs were much higher than the external benefits. Over four decades, 269,000 hectares (660,000 acres) of Indonesian mangroves have been converted to shrimp farms. Most of these farms are abandoned within a decade because of the toxin build-up and nutrient loss.

Salmon farms are typically sited in pristine coastal ecosystems which they then pollute. A farm with 200,000 salmon discharges more fecal waste than a city of 60,000 people.

This waste is discharged directly into the surrounding aquatic environment, untreated, often containing antibiotics and pesticides." There is also an accumulation of heavy metals on the benthos (seafloor) near the salmon farms, particularly copper andzinc.

### Genetic Modification

Salmon have been genetically modified for faster growth, although they are not approved for commercial use, in the face of opposition. One study, in a laboratory setting, found that modified salmon mixed with their wild relatives were aggressive in competing, but ultimately failed.

### Animal Welfare

As with the farming of terrestrial animals, social attitudes influence the need for humane practices and regulations in farmed marine animals. Under the guidelines advised by the Farm Animal Welfare Council good animal welfare means both fitness and a sense of well being in the animal's physical and mental state. This can be defined by the Five Freedoms:

- Freedom from hunger & thirst
- Freedom from discomfort
- Freedom from pain, disease, or injury
- Freedom to express normal behaviour
- Freedom from fear and distress

However, the controversial issue in aquaculture is whether fish and farmed marine invertebrates are actually sentient, or have the perception and awareness to experience suffering.

Although no evidence of this has been found in marine invertebrates, recent studies conclude that fish do have the necessary receptors (nociceptors) to sense noxious stimuli and so are likely to experience states of pain, fear and stress. Consequently, welfare in aquaculture is directed at vertebrates; finfish in particular.

### Common Welfare Concerns

Welfare in aquaculture can be impacted by a number of issues such as stocking densities, behavioural interactions, disease and parasitism.

A major problem in determining the cause of impaired welfare is that these issues are often all interrelated and influence each other at different times.

Optimal stocking density is often defined by the carrying capacity of the stocked environment and the amount of individual space needed by the fish, which is very species specific.

Although behavioural interactions such as shoaling may mean that high stocking densities are beneficial to some species, in many cultured species high stocking densities may be of concern. Crowding can constrain normal swimming behaviour, as well as increase aggressive and competitive behaviours such as cannibalism, feed competition, territoriality and dominance/subordination hierarchies.

This potentially increases the risk of tissue damage due to abrasion from fish-to-fish contact or fish-to-cage contact.

Fish can suffer reductions in food intake and food conversion efficiency. In addition, high stocking densities can result in water flow being insufficient, creating inadequate oxygen supply and waste product removal. Dissolved oxygen is essential for fish respiration and concentrations below critical levels can induce stress and even lead to asphyxiation. Ammonia, a nitrogen excretion product, is highly toxic to fish at accumulated levels, particularly when oxygen concentrations are low.

Many of these interactions and effects cause stress in the fish, which can be a major factor in facilitating fish disease. For many parasites, infestation depends on the host's degree of mobility, the density of the host population and vulnerability of the host's defence system. Sea lice are the primary parasitic problem for finfish in aquaculture, high numbers causing widespread skin erosion and haemorrhaging, gill congestion, and increased mucus production. There are also a number of prominent viral and bacterial pathogens that can have severe effects on internal organs and nervous systems.

## Improving Welfare

The key to improving welfare of marine cultured organisms is to reduce stress to a minimum, as prolonged or repeated stress can cause a range of adverse effects. Attempts to minimise stress can occur throughout the culture process. During grow out it is important to keep stocking densities at appropriate levels specific to each species, as well as separating size classes and grading to reduce aggressive behavioural interactions. Keeping nets and cages clean can assist positive water flow to reduce the risk of water degradation.

Not surprisingly disease and parasitism can have a major effect on fish welfare and it is important for farmers not only to manage infected stock but also to apply disease prevention measures. However, prevention methods, such as vaccination, can also induce stress because of the extra handling and injection. Other methods include adding antibiotics to feed, adding chemicals into water for treatment baths and biological control, such as using cleaner wrasse to remove lice from farmed salmon.

Many steps are involved in transport, including capture, food deprivation to reduce faecal contamination of transport water, transfer to transport vehicle via nets or pumps, plus transport and transfer to the delivery location. During transport water needs to be maintained to a high quality, with regulated temperature, sufficient oxygen and minimal waste products. In some cases anaesthetics may be used in small doses to calm fish before transport.

Aquaculture is sometimes part of an environmental rehabilitation program or as an aid in conserving endangered species.

## Prospects

Global wild fisheries are in decline, with valuable habitat such as estuaries in critical condition. The aquaculture or farming of piscivorous fish, like salmon, does not help the problem because they need to eat products from other fish, such as fish meal and fish oil. Studies have shown that salmon farming has major negative impacts on wild salmon, as well as the forage fish that need to be caught to feed them. Fish that are higher on the food chain are less efficient sources of food energy.

Apart from fish and shrimp, some aquaculture undertakings, such as seaweed and filter-feeding bivalve mollusks like oysters, clams, mussels and scallops, are relatively benign and even environmentally restorative. Filter-feeders filter pollutants as well as nutrients from the water, improving water quality. Seaweeds extract nutrients such as inorganic nitrogen and phosphorus directly from the water, and filter-feeding mollukscan extract nutrients as they feed on particulates, such as phytoplankton and detritus.

Some profitable aquaculture cooperatives promote sustainable practices. New methods lessen the risk of biological and chemical pollution through minimising fish stress, fallowing netpens, and applying Integrated Pest Management. Vaccines are being used more and more to reduce antibiotic use for disease control.

Onshore recirculating aquaculture systems, facilities using polyculture techniques, and properly sited facilities (for example, offshore areas with strong currents) are examples of ways to manage negative environmental effects.

Recirculating aquaculture systems (RAS) recycle water by circulating it through filters to remove fish waste and food and then recirculating it back into the tanks. This saves water and the waste gathered can be used incompost or, in some cases, could even be treated and used on land.

While RAS was developed with freshwater fish in mind, scientist associated with the Agricultural Research Service have found a way to rear saltwater fish using RAS in low-salinity waters. Although saltwater fish are raised in off-shore cages or caught with nets in water that typically has a salinity of 35 parts per thousand (ppt), scientists were able to produce healthy pompano, a saltwater fish, in tanks with a salinity of only 5 ppt. Commercialising low-salinity RAS are predicted to have positive environmental and economical effects.

Unwanted nutrients from the fish food would not be added to the ocean and the risk of transmitting diseases between wild and farm-raised fish would greatly be reduced.

The price of expensive saltwater fish, such as the pompano and combia used in the experiments, would be reduced. However, before any of this can be done researchers must study every aspect of the fish's lifecycle, including the amount of ammonia and nitrate the fish will tolerate in the water, what to feed the fish during each stage of its lifecycle, the stocking rate that will produce the healthiest fish, etc.

## Aquaponics

***Figure:*** *A small, portable aquaponics system*

Aquaponics is a sustainable food production system that combines a traditional aquaculture (raising aquatic animals such as snails, fish, crayfish or prawns in tanks) with hydroponics (cultivating plants in water) in a symbiotic environment. In the aquaculture, effluents accumulate in the water, increasing toxicity for the fish.

This water is led to a hydroponic system where the by-products from the aquaculture are filtered out by the plants as vital nutrients, after which the cleansed water is recirculated back to the animals. The term *aquaponics* is a portmanteau of the terms *aquaculture* and *hydroponic*.

Aquaponic systems vary in size from small indoor or outdoor units to large commercial units, using the same technology. The systems usually contain fresh water, but salt water systems are plausible depending on the type of aquatic animal and which plants. Aquaponic science may still be considered to be at an early stage.

### Function

Aquaponics consists of two main parts, with the aquaculture part for raising aquatic animals and the hydroponics part for growing plants. Aquatic effluents resulting from uneaten feed or raising animals like fish, accumulates in water due to the closed system recirculation of most aquaculture systems. The effluent-rich water becomes toxic to the aquatic animal in high concentrations but these effluents are nutrients essential for plant growth. Although consisting primarily of these two parts, aquaponics systems are usually grouped into several components or subsystems responsible for the effective removal of solid wastes, for adding bases to neutralise acids, or for maintaining water oxygenation. Typical components include:

- *Rearing tank*: the tanks for raising and feeding the fish;
- *Solids removal*: a unit for catching uneaten food and detached biofilms, and for settling out fine particulates;
- *Biofilter*: a place where the nitrification bacteria can grow and convert ammonia into nitrates, which are usable by the plants;
- *Hydroponics subsystem*: the portion of the system where plants are grown by absorbing excess nutrients from the water;
- *Sump*: the lowest point in the system where the water flows to and from which it is pumped back to the rearing tanks.

Depending on the sophistication and cost of the aquaponics system, the units for solids removal, biofiltration, and/or the hydroponics subsystem may be combined into one unit or subsystem, which prevents the water from flowing directly from the aquaculture part of the system to the hydroponics part.

### Nitrification

Nitrification, the aerobic conversion of ammonia into nitrates, is one of the most important functions in an aquaponics system as it reduces the toxicity of the water for fish, and allows the resulting nitrate compounds to be removed by the plants for nourishment.

Ammonia is steadily released into the water through the excreta and gills of fish as a product of their metabolism, but must be filtered out of the water since higher concentrations of ammonia (commonly between 0.5 and 1 ppm) can kill fish. Although plants can absorb ammonia from the water to some degree, nitrates are assimilated more easily, thereby efficiently reducing the toxicity of the water for fish. Ammonia can be converted into other nitrogenous compounds through healthy populations of:

- *Nitrosomonas*: bacteria that convert ammonia into nitrites, and
- *Nitrobacter*: bacteria that convert nitrites into nitrates.

In an aquaponics system, the bacteria responsible for this process form a biofilm on all solid surfaces throughout the system that are in constant contact with the water. The submerged roots of the vegetables combined have a large surface area, so that many bacteria can accumulate there. Together with the saliency of ammonia and nitrites in the water, the surface area determines the speed with which nitrification takes place. Care for these bacterial colonies is important as to regulate the full assimilation of ammonia and nitrite. This is why most aquaponics systems include a biofiltering unit, which helps facilitate growth of these microorganisms.

Typically, after a system has stabilised ammonia levels range from 0.25 to 2.0 ppm; nitrite levels range from 0.25 to 1 ppm, and nitrate levels range from 2 to 150 ppm. During system startup, spikes may occur in the levels of ammonia (up to 6.0 ppm) and nitrite (up to 15 ppm), with nitrate levels peaking later in the startup phase. Since the nitrification process acidifies the water, non-sodium bases such as potassium hydroxide or calcium hydroxide can be added for neutralising the water's pH if insufficient quantities are naturally present in the water to provide a buffer against acidification. In addition, selected minerals or nutrients such as iron can be added in addition to the fish waste that serves as the main source of nutrients to plants.

A good way to deal with solids buildup in aquaponics is the use of worms, which liquefy the solid organic matter so that it can be utilised by the plants and/or animals.

### Hydroponics Subsystem

Plants are grown as in hydroponics systems, with their roots immersed in the nutrient-rich effluent water. This enables them to filter out the ammonia that is toxic to the aquatic animals, or its metabolites. After the water has passed through the hydroponic subsystem, it is cleaned and oxygenated, and can return to the aquaculture vessels. This cycle is continuous. Common aquaponic applications of hydroponic systems include:

- *Deep-water raft aquaponics*: styrofoam rafts floating in a relatively deep aquaculture basin in troughs.
- *Recirculating aquaponics*: solid media such as gravel or clay beads, held in a container that is flooded with water from the aquaculture. This type of aquaponics is also known as *closed-loop aquaponics*.

- *Reciprocating aquaponics*: solid media in a container that is alternately flooded and drained utilising different types of siphon drains. This type of aquaponics is also known as *flood-and-drain aquaponics* or *ebb-and-flow aquaponics*.
- Other systems use towers that are trickle-fed from the top, nutrient film technique channels, horizontal PVC pipes with holes for the pots, plastic barrels cut in half with gravel or rafts in them. Each approach has its own benefits.

Most green leaf vegetables grow well in the hydroponic subsystem, although most profitable are varieties of chinese cabbage, lettuce, basil, roses, tomatoes, okra, cantaloupe and bell peppers. Other species of vegetables that grow well in an aquaponic system include beans, peas, kohlrabi, watercress, taro, radishes, strawberries, melons, onions, turnips, parsnips, sweet potato and herbs. Since plants at different growth stages require different amounts of minerals and nutrients, plant harvesting is staggered with seedings growing at the same time as mature plants.

This ensures stable nutrient content in the water because of continuous symbiotic cleansing of toxins from the water.

### Aquaculture Subsystem

Freshwater fish are the most common aquatic animal raised using aquaponics, although freshwater crayfish and prawns may also be used. In practice, tilapia are the most popular fish for home and commercial projects that are intended to raise edible fish, although barramundi, Silver Perch, Eel-tailed catfish or tandanus catfish, Jade perch and Murray cod are also used.

For temperate climates when there isn't ability or desire to maintain water temperature, bluegill and catfish are suitable fish species for home systems. Koi and goldfish may also be used, if the fish in the system need not be edible.

### Normal Operations

Aquaponic systems do not typically discharge or exchange water under normal operation, but instead recirculate and reuse water very effectively. The system relies on the relationship between the animals and the plants to maintain a stable aquatic environment that experience a minimum of fluctuation in ambient nutrient and oxygen levels. Water is only added to replace water loss from absorption and transpiration by plants, evapouration into the air from surface water, overflow from the system from rainfall, and removal of biomass such as settled solid wastes from the system. As a result, aquaponics uses approximately 2% of the water that a conventionally irrigated farm requires for the same vegetable production. This allows for aquaponic production of both crops and fish in areas where water or fertile land is scarce.

Aquaponic systems can also be used to replicate controlled wetland conditions that are useful for water treatment by reclaiming potable water from typical household sewage.

The nutrient-filled overflow water can be accumulated in catchment tanks, and reused to accelerate growth of crops planted in soil, or it may be pumped back into the aquaponic system to top up the water level..

The three main inputs to the system are water, feed given to the aquatic animals, and electricity to pump water between the aquaculture subsystem and the hydroponics subsystem. Spawn or fry may be added to replace grown fish that are taken out from the system to retain a stable system. In terms of outputs, an aquaponics system may continually yield plants such as vegetables grown in hydroponics, and edible aquatic species raised in an aquaculture.

### History

### Ancient

Aquaponics has ancient roots, although there is some debate on its first occurrence:

- Aztec cultivated agricultural islands known as *chinampas* and are considered by some as the first form of aquaponics for agricultural use where plants were raised on stationary (and sometime movable) islands in lake shallows and waste materials dredged from the Chinampa canals and surrounding cities are used to manually irrigate the plants.
- South China and Thailand who cultivated and farmed rice in paddy fields in combination with fish are cited as examples of early aquaponics. These polycultural farming systems existed in many Far Eastern countries and raised fish such as the oriental loach swamp eel, Common and crucian carp as well as pond snails in the paddies.

## Regions

### North America

***United States:*** While the development of aquaponics is often attributed to the various works of the New Alchemy Institute and the works of Dr. Mark McMurtry et al. at the North Carolina State University, many papers of initial development of aquaponics concepts pre-date both institutions by nearly a decade.

Tom and Paula Speraneo, owners of a small greenhouse operation near West Plains, Missouri, modified the North Carolina State method and raised tilapia in above-ground tanks inside a solar greenhouse. The effluent from the tanks was used to fertigate gravel-cultured vegetables in raised benches.

In addition, the Speraneos manipulated the watering cycle, and this methodology of the Speranos forms the basis for the style of "flood & drain" media grow bed aquaponics systems that have been widely adopted in Australia based on models promoted by Joel Malcolm and Murray Hullam, and are now gaining increasing popularity in the United States.

Inspired by the successes of the New Alchemy Institute and the North Carolina State University with aquaponics, other institutes followed suit. Besides the reciprocating aquaponics based on the techniques developed by Dr. Mark McMurtry et al. at the North Carolina State University, Dr. James Rakocy and his colleagues at the University of the Virgin Islands researched and developed the "Deep Water" or "Raft Culture" aquaponics The system combines tilapia with various vegetables.

In 1997 Rebecca L. Nelson and John S. Pade began publishing the Aquaponics Journal, a quarterly scientific journal that brings together research and various applications of aquaponics from around the globe. In 2008, they wrote and published the first book on aquaponics, *Aquaponic Food Production.* Since then there have been numerous books, and "e-books" appear concerning aquaponics.

In 2010 the Aquaponic Gardening Community was started, which has now become the largest on-line gathering place for aquaponics enthusiasts in North America. Members of that community launched the Aquaponics Association in 2011 after the first Aquaponics Conference was held in Orlando, Florida in September of that year.

Recent years have seen a shift towards community integration of aquaponics, such as the nonprofit foundation Growing Power that offers Milwaukee youth job opportunities and training while growing food for their community. The model has spawned several satellite projects in other cities, such as New Orleans where the Vietnamese fisherman community has suffered from the Horizon Disaster. In addition, aquaponic gardeners from all around the world have gathered in online community sites and forums to openly share their experiences and move the development of this fantastic form of gardening forward.

### Canada

The first aquaponics research in Canada was a small system added onto existing aquaculture research at a research station in Lethbridge. Canada saw a rise in aquaponics setups throughout the 90s, predominantly as commercial installations, that for example combine trout with floating lettuce production, or to water fruiting vegetable crops that warm up the water too much to be recirculated back into the fish ponds.

Eels are also known to be raised. A set-up based on the deep water system developed at the University of Virgin Islands was built in a greenhouse at Brooks, Alberta where Dr. Nick Savidov and colleagues researched aquaponics from a background of plant science.

The team made findings on rapid root growth in aquaponics systems, on closing the solid waste loop, and that because of certain advantages in the system over traditional aquaculture, the system can run well at a low pH level, which is favoured by plants but not fish.

The Edmonton Aquaponics Society in Northern Alberta is adapting Dr. Savidov's commercially sized system to a smaller scale prototype that can be operated by families, small groups, or restaurants. They intend to further develop the closed solid waste loop.

### South America

### Barbados

Barbados is a densely populated island that deals with water scarcity. In 40 years' time, focus has shifted from domestic fruit and vegetable production on small farms to importing 80% of all fruits and vegetables for cost reasons. Aquaponics would make Barbados and other Caribbean islands less dependent on the world food market and reduce stress on the dwindling fish supplies. An inter-organisational project that started in late 2009 sets out to encourage and enable Barbadians to start aquaponics at home, with revenue generated by selling produce to tourists.

### Asia

### Bangladesh

Bangladesh is the world's most densely populated country. Most of its farmers use/misuse agrochemicals to enhance food production/ storage life. However, the country is yet to establish a good monitoring system to ensure chemical-residue free safe food in the market, capable city dwellers are looking for a better way to grow their own safe foods for domestic consumption. To help support mainly this client group and people living in adverse climatic conditions such as salinity-prone southern part and flood-pronehaor area in the eastern region, a low-cost model of backyard aqaponics system has been developed by Professor Dr. M.A. Salam at the Department of Aquaculture of Bangladesh Agricultural University, Mymensingh. The initial model has been developed in 2011, and demonstrated and reported to the press in January 2012. Further researches may be necessary to develop appropriate models for different situations among client groups.

### Taiwan

Taiwan is a densely populated island that is faced with freshwater scarcity. Dispensation of water is government-controlled. The closed-loop system of aquaponics is used by agricultural farmers to save water by also rearing fish, while fish farmers grow plants that filter the water from the fish tanks. Taiwan Aquaponics Association is the not-for profit federation to promote Aquaponics at Taiwan. 27

### Australia

Tom and Paula Speraneo, owners of a small greenhouse operation near West Plains, Missouri, modified the North Carolina State method developed by Dr. Mark McMurtry et al., and raised tilapia in above-ground tanks inside a solar greenhouse. The effluent from the tanks was used to fertigate gravel-cultured vegetables in raised benches. In addition, the Speraneos manipulated the watering cycle, and this methodology of the Speranos forms the basis for the style of "flood & drain" media grow bed aquaponics systems that have been widely adopted in Australia based on models promoted by Joel Malcolm and

Murray Hullam, and are now gaining increasing popularity in the United States. Due to a ban on tilapia in all states except for Western Australia, native freshwater fish including Silver Perch, Jade Perch, Sleepy Cod, Murray Cod and Barramundi are popular in aquaponics and aquaculture systems, along with non-native Rainbow Trout, Brown Trout and Crayfish such as common Yabby and Redclaw.

### Pros and Cons

The unique advantages of aquaponic systems are:

- Conservation through constant water reuse and recycling.
- Organic fertilization of plants with natural fish emulsion.
- The elimination of solid waste disposal from intensive aquaculture.
- The reduction of needed cropland to produce crops.
- The overall reduction of the environmental footprint of crop production.
- Building small efficient commercial installations near markets reduces food miles.
- Reduction of pathogens that often plague aquaculture production systems.

Some conceivable disadvantages with aquaponics are:

- Initial expenses for housing, tank, plumbing, pumps, and grow beds.
- The infinite number of ways in which a system can be configured lends itself to equally varying results, conflicting research, and successes or failures.
- Some aquaponic installations rely heavily on man-made energy, technological solutions, and environmental control to achieve recirculation and water/ambient temperatures. However, if a system is designed with energy conservation in mind, using alternative energy and a reduced number of pumps by letting the water flow downwards as much as possible, it can be highly energy efficient.
- While careful design can minimise the risk, aquaponics systems can have multiple 'single points of failure' where problems such as an electrical failure or a pipe blockage can lead to a complete loss of fish stock.

- Like all aquaculture based systems, stock feed usually consists of fish meal derived from lower value species. Ongoing depletion of wild fish stocks makes this practice unsustainable. Organic fish feeds may prove to be a viable alternative that negates this concern. Other alternatives include growing duckweed with an aquaponics system that feeds the same fish grown on the system, excess worms grown from vermiculture composting, as well as growing black soldier fly larvae to feed to the fish using composting grub growers.

## Copper Alloys in Aquaculture

Recently, copper alloys have become important netting materials in aquaculture (the farming of aquatic organisms including fish farming). Various other materials including nylon, polyester, polypropylene, polyethylene, plastic-coated welded wire, rubber, patented twine products (Spectra, Dyneema), and galvanised steelare also used for netting in aquaculture fish enclosures around the world. All of these materials are selected for a variety of reasons, including design feasibility, material strength, cost, and corrosion resistance.

What sets copper alloys apart from the other materials used in fish farming is that copper alloys are antimicrobial, that is, they destroy bacteria, viruses, fungi, algae, and other microbes.

In the marine environment, the antimicrobial/algaecidal properties of copper alloys prevent biofouling, which can briefly be described as the undesirable accumulation, adhesion, and growth of microorganisms, plants, algae, tubeworms, barnacles, mollusks, and other organisms on man-made marine structures. By inhibiting microbial growth, copper alloy aquaculture pens avoid the need for costly net changes that are necessary with other materials. The resistance of organism growth on copper alloy nets also provides a cleaner and healthier environment for farmed fish to grow and thrive.

In addition to their antifouling benefits, copper alloys have strong structural and corrosion-resistant properties in marine environments.

It is the combination of all of these properties – antifouling, high strength, and corrosion resistance – that has made copper alloys a desirable material for such marine applications as condenser tubing, water intake screens, ship hulls, offshore structure, and sheathing. In the past 25 years or so, the benefits of copper alloys have caught the attention of the marine aquaculture industry. The industry is now actively deploying copper alloy netting and structural materials in commercial large-scale fish farming operations around the world.

### Importance of Aquaculture

Much has been written about the degradation and depletion of natural fish stocks in rivers, estuaries, and the oceans ( Overfishing). Because industrial fishing has become extremely efficient, ocean stocks of large fish, such as tuna, cod, and halibut have declined by 90% in the past 50 years.

Aquaculture, an industry that has emerged only in recent decades, has become one of the fastest growing sectors of the world food economy. Aquaculture already supplies more than half of the world's demand for fish. This percentage is predicted to increase dramatically over the next few decades.

### The Problem of Biofouling

Biofouling is one of the most important problems in aquaculture. Biofouling occurs on non-copper materials in the marine environment, including fish pen surfaces and nettings. For example, it was noted that the open area of a mesh immersed for only seven days in a Tasmanian aquaculture operation decreased by 37% as a result of biofouling.

The biofouling process begins when algae spores, marine invertebrate larvae, and other organic material adhere to surfaces submerged in marine environments (e.g., fish nets in aquaculture). Bacteria then encourage the attachment of secondary unwanted colonisers.

Biofouling has strong negative impacts on aquaculture operations. Water flow and dissolved oxygen are inhibited due to clogged nets in fish pens. The end result is often diseased fish from infections, such as netpen liver disease, amoebic gill disease, and parasites. Other negative impacts include increased fish mortalities, decreased fish growth rates, premature fish harvesting, reduced fish product values and profitability, and an adversely impacted environment near fish farms.

Biofouling adds enormous weight to submerged fish netting. Two hundredfold increases in weight have been reported. This translates, for example, to two thousand pounds of unwanted organisms adhered to what was once a clean 10-pound fish pen net. In South Australia, biofouling weighing 6.5 tonnes (approximately 13,000 pounds) was observed on a fish pen net. This extra burden often results in net breakage and additional maintenance costs.

To combat parasites from biofouling in finfish aquaculture, treatment protocols such as cypermethrin, azamethiphos, and emamectin benzoate may be administered, but these have been found to have detrimental environmental effects, for example, in lobster operations.

To treat diseases in fish raised in biofouled nets, fish stocks are administered antibiotics. The antibiotics can have unwanted long-term health effects on consumers and on coastal environments near aquaculture operations.

To combat biofouling, operators often implement costly maintenance measures, such as frequent net changing, cleaning/ removal of unwanted organisms from nets, net repairs, and chemical treatment including antimicrobial coatings on nylon nets.

The cost of antifouling a single salmon net can be several thousand British pounds. In some sectors of the European aquaculture industry, cleaning biofouled fish and shellfish pens can cost 5–20% of its market value.

Heavy fouling can reduce the saleable product in nets by 60–90%. Antifouling coatings are often used on nylon nets because the process is more economical than manual cleaning. When nylon nets are coated with antifouling compounds, the coatings repel biofouling for a period of time, usually between several weeks to several months. However, the nets eventually succumb to biofouling.

Antifouling coatings containing cuprous oxide algaecide/biocide are the coatings technology used almost exclusively in the fish farming industry today. The treatments usually flake off within a few weeks to six to eight months.

Biofouled nets are replaced after several months of service, depending on environmental conditions, in a complicated, costly, and labour-intensive operation that involves divers and specialised personnel. During this process, live fish in nets must be transferred to clean pens, which causes undue stress and asphyxiation that results in some loss of fish. Biofouled nets that can be reused are washed on land via manual brushing and scrubbing or high-pressure water hosing. They are then dried and re-impregnated with antifouling coatings.

A line of net cleaners is available for in-situ washings where permitted. But, even where not permitted by environmental, fisheries, maritime, and sanitary authorities, should the lack of dissolved oxygen in submerged pens create an emergency condition that endangers the health of fish, divers may be deployed with special in situ cleaning machinery to scrub biofouled nets.

The aquaculture industry is addressing the negative environmental impacts from its operations. As the industry evolves, a cleaner, more sustainable aquaculture industry is expected to emerge, one that may increasingly rely on materials with anti-fouling, anti-corrosive, and strong structural properties, such as copper alloys.

### *Antifouling Properties of Copper Alloys*

In the aquaculture industry, sound animal husbandry translates to keeping fish clean, well fed, healthy, and not overcrowded. One solution to keeping farmed fish healthy is to contain them in antifouling copper alloy nets and structures.

Researchers have attributed copper's resistance to biofouling, even in temperate waters, to two possible mechanisms:

1) a retarding sequence of colonisation through release of antimicrobial copper ions, thereby preventing the attachment of microbial layers to marine surfaces; and,

2) separating layers that contain corrosive products and the spores of juveniles or macro-encrusting organisms.

The most important requirement for optimum biofouling resistance is that the copper alloys should be freely exposed or electrically insulated from less noble alloys and fromcathodic protection. Galvanic coupling to less noble alloys and cathodic protection prevent copper ion releases from surface films and therefore reduce biofouling resistance.

As temperatures increase and water velocities decrease in marine waters, biofouling rates dramatically rise. However, copper's resistance to biofouling is observed even in temperate waters. Studies in La Herradura Bay, Coquimbo, Chile, where biofouling conditions are extreme, demonstrated that a copper alloy (90% copper, 10% nickel) avoided macro-encrusting organisms.

### Corrosion Behaviour of Copper Alloys

Copper alloys used in sea water service have low general corrosion rates but also have a high resistance to many localised forms of corrosion. A technical discussion regarding various types of corrosion, application considerations (e.g., depth of installations, effect of polluted waters, sea conditions), and the corrosion characteristics of several copper alloys used in aquaculture netting is available (i.e., copper-nickel, copper-zinc, and copper-silicon).

### Environmental Performance of Copper Alloy Mesh

Many complicated factors influence the environmental performance of copper alloys in aquaculture operations. A technical description of antibiofouling mechanisms, fish health and welfare, fish losses due to escapes and predator attacks, and reduced life cycle environmental impacts is summarised in this reference.

### Types of Copper Alloys

Copper-zinc brass alloys are currently (2011) being deployed in commercial-scale aquaculture operations in Asia, South America and the USA (Hawaii). Extensive research, including demonstrations and trials, are currently being implemented on two other copper alloys: copper-nickel and copper-silicon.

Each of these alloy types has an inherent ability to reduce biofouling, pen waste, disease, and the need for antibiotics while simultaneously maintaining water circulation and oxygen requirements. Other types of copper alloys are also being considered for research and development in aquaculture operations.

The University of New Hampshire is in the midst of conducting experiments under the auspices of the International Copper Association (ICA) to evaluate the structural, hydrodynamic, and antifouling response of copper alloy nets.

Factors to be determined from these experiments, such as drag, pen dynamic loads, material loss, and biological growth – well documented for nylon netting but not fully understood for copper-nickel alloy nets – will help to design fish pen enclosures made from these alloys.

The East China Sea Fisheries Research Institute, in Shanghai, China, is also conducting experimental investigations on copper alloys for ICA.

### Copper-zinc Alloys

The Mitsubishi-Shindoh Co., Ltd., has developed a proprietary copper-zinc brass alloy, called UR30, specifically designed for aquaculture operations. The alloy, which is composed of 64% copper, 35.1% zinc, 0.6% tin, and 0.3% nickel, resists mechanical abrasion when formed into wires and fabricated into chain link, woven, or other types of flexible mesh. Corrosion rates depend on the depth of submersion and seawater conditions. The average reported corrosion rate reported for the alloy is <5 ìm/yr based on 2- and 5-year exposure trials in seawater.

The Ashimori Industry Company, Ltd., has installed approximately 300 flexible pens with woven chain link UR30 meshes in Japan to raise Seriola (i.e., yellowtail, amberjack, kingfish, hamachi).

The company has installed another 32 brass pens to raise Atlantic salmon at the Van Diemen Aquaculture operations in Tasmania, Australia. In Chile, EcoSea Farming S.A. has installed a total of 62 woven chain link brass mesh pens to raise troutand Atlantic salmon. In Panama, China, Korea, Turkey, and the USA, demonstrations and trials are underway using flexible pens with woven chain link UR30 and other mesh forms and a range of copper alloys.

To date, in over 10 years of aquaculture experience, chain link mesh fabricated by these brass alloys have not suffered from dezincification, stress corrosion cracking, or erosion corrosion.

### Copper-nickel Alloys

Copper-nickel alloys were developed specifically for seawater applications over five decades ago. Today, these alloys are being investigated for their potential use in aquaculture.

Copper-nickel alloys for marine applications are usually 90% copper, 10% nickel, and small amounts of manganese and iron to enhance corrosion resistance.

The seawater corrosion resistance of copper-nickel alloys results in a thin, adherent, protective surface film which forms naturally and quickly on the metal upon exposure to clean seawater.

The rate of corrosion protective formation is temperature dependent. For example, at 27 °C (i.e., a common inlet temperature in the Middle East), rapid film formation and good corrosion protection can be expected within a few hours. At 16 °C, it could take 2–3 months for the protection to mature. But once a good surface film forms, corrosion rates decrease, normally to 0.02–0.002 mm/yr, as protective layers develop over a period of years. These alloys have good resistance to chloride pitting and crevice corrosion and are not susceptible to chloride stress corrosion.

### Copper-silicon Alloys

Copper-silicon has a long history of use as screws, nuts, bolts, washers, pins, lag lag bolts, and staples in wooden sailing vessels in marine environments. The alloys are often composed of copper, silicon, and manganese. The inclusion of silicon strengthens the metal.

As with the copper-nickel alloys, corrosion resistance of copper-silicon is due to protective films that form on the surface over a period of time. General corrosion rates of 0.025–0.050mm have been observed in quiet waters.

This rate decreases towards the lower end of the range over long-term exposures (e.g., 400–600 days). There is generally no pitting with the silicon-bronzes.

Also there is good resistance to erosion corrosion up to moderate flow rates. Because copper-silicon is weldable, rigid pens can be constructed with this material.

Also, because welded copper-silicon mesh is lighter than copper-zinc chain link, aquaculture enclosures made with copper-silicon may be lighter in weight and therefore a potentially less expensive alternative.

Luvata Appleton, LLC, is researching and developing a line of copper alloy woven and welded meshes, including a patent-pending copper silicon alloy, that are marketed under the trade name Seawire. Copper-silicon alloy meshes have been developed by the firm to raise various marine organisms in test trials that are now in various stages of evaluation.

These include raising cobia in Panama, lobsters in Maine, and crabs in the Chesapeake Bay. The company is working with various universities to study its material, including the University of Arizona to study shrimp, the University of New Hampshire to study cod, and Oregon State Universityto study oysters. Results will be provided when the studies are concluded.

## Fisheries Science

***Figure:*** *Fisheries scientists sorting a catch of small fish and Norway lobster*

Fisheries science is the academic discipline of managing and understanding fisheries.

It is a multidisciplinary science, which draws on the disciplines of limnology, oceanography, freshwater biology, marine biology, conservation, ecology, population dynamics, economics and management to attempt to provide an integrated picture of fisheries.

In some cases new disciplines have emerged, as in the case of bioeconomics.Fisheries science is typically taught in a university setting, and can be the focus of an undergraduate, master's or Ph.D. program. Some universities offer fully integrated programs in fisheries science.

### *Notable Contributors*

- Spencer F. Baird – founding scientist of the United States Fish Commission
- Fedor I. Baranov - Russian scientist and the father of the *Baranov catch equation*, thereby being one of the founding fathers of fisheries science
- Ludwig von Bertalanffy – Austrian-born biologist and a founder of general systems theory
- Ray Beverton – English fisheries biologist; known for the Beverton–Holt model (with Sidney Holt), cred with being one of the founders of fisheries science

- Villy Christensen – fisheries scientist and ecosystem modeler, known for his work on the development of Ecopath
- John N. Cobb – founder of the first college of fisheries in the United States, the University of Washington College of Fisheries, in 1919
- David Cushing – English born fisheries biologist, who is cred with the development of the match/mismatch hypothesis
- W. Harry Everhart – American fisheries scientist, educator, administrator, and author of several widely used fisheries texts
- Rainer Froese – Known for his work on the development and coordination of FishBase
- Gotthilf Hempel – German marine biologist and oceanographer, and co-founder of the Alfred Wegener Institute for Polar and Marine Research
- Walther Herwig – Prussian lawyer and promoter of high seas fishing and research
- Ray Hilborn – Canadian-born fisheries biologist, with strong contributions towards fisheries management
- Johan Hjort – Norwegian fisheries biologist, marine zoologist, and oceanographer
- Bruno Hofer – German fishery scientist, cred with being the founder of fish pathology
- Sidney Holt – English fisheries biologist; known for the Beverton–Holt model (with Ray Beverton), cred with being one of the founders of fisheries science
- Uwe Kils – German marine biologist specialising in planktology. Inventor of the ecoSCOPE
- Robert T. Lackey – Canadian born fisheries scientist and political scientist known for his work involving the role of science in policy making
- Leo Margolis – Canadian parasitologist and head of the Pacific Biological Station in Nanaimo, British Columbia
- R. J. McKay – Australian-born biologist and a specialist in translocated freshwater fishes
- Ransom A. Myers – Canadian marine biologist and conservationist

- Daniel Pauly – prominent French-born fisheries scientist, known for his work studying human impacts on global fisheries
- Tony J. Pitcher – known for work on the impacts of fishing, management appraisals and the shoaling behaviour of fish
- Michael A. Rice – American known for work on molluscan fisheries
- Bill Ricker – Canadian fisheries biologist, known for the Ricker model, cred with being one of the founders of fisheries science
- Ed Ricketts – a colourful American marine biologist and philosopher who introduced ecology to fisheries science.
- Callum Roberts – British marine conservation biologist, known for his work on the role marine reserves play in protecting marine ecosystems
- Harald Rosenthal – German hydrobiologist known for his work in fish farming and ecology
- Carl Safina – author of several writings on marine ecology and the ocean
- Georg Ossian Sars – Norwegian marine biologist cred with the discovery of a number of new species and known for his analysis of cod fisheries
- Tore Schweder – Norwegian statistician whose work includes the assessment of marine resources
- Milner Baily Schaefer – notable for his work on the population dynamics of fisheries
- Ussif Rashid Sumaila – noted for his analysis of the economic aspects of fisheries
- Fred Utter – noted as the founding father of the field of fishery genetics
- Carl Walters – American born biologist known for his work involving fisheries stock assessments, the adaptive management concept, and ecosystem modelling

## Fish Hatchery

A fish hatchery is a "place for artificial breeding, hatching and rearing through the early life stages of animals, finfish and shellfish in particular". Hatcheries produce larval and juvenile fish (and shellfish and crustaceans) primarily to support the aquaculture industry where they are transferred to on-growing systems i.e. fish farms to reach harvest size.

Some species that are commonly raised in hatcheries include Pacific oysters, shrimp, Indian prawns, salmon, tilapia andscallops. The value of global aquaculture production is estimated to be US$98.4 billion in 2008 with China significantly dominating the market, however the value of aquaculture hatchery and nursery production has yet to be estimated.

Additional hatchery production for small-scale domestic uses, which is particularly prevalent in South-East Asia or for conservation programmes, has also yet to be quantified.

There is much interest in supplementing exploited stocks of fish by releasing juveniles that may be wild caught and reared in nurseries before transplanting, or produced solely within a hatchery.

Culture of finfish larvae has been utilised extensively in the United States in stock enhancement efforts to replenish natural populations. The U.S. Fish and Wildlife Service have established a National Fish Hatchery System to support the conservation of native fish species.

### Purpose

Hatcheries produce larval and juvenile fish and shellfish for transferal to aquaculture facilities where they are 'on-grown' to reach harvest size. Hatchery production confers three main benefits to the industry;

1. *Out of season production:* Consistent supply of fish from aquaculture facilities is an important market requirement. Broodstock conditioning can extend the natural spawning season and thus the supply of juveniles to farms. Supply can be further guaranteed by sourcing from hatcheries in the opposite hemisphere i.e. with opposite seasons.
2. *Genetic improvement:* Genetic modification is conducted in some hatcheries to improve the quality and yield of farmed species. Artificial fertilisation facilitates selective breeding programs which aim to improve production characteristics such as growth rate, disease resistance, survival, colour,

increased fecundity and/or lower age of maturation. Genetic improvement can be mediated by selective breeding, via hybridisation, or other genetic manipulation techniques.

3. *Reduce dependence on wild-caught juveniles:* In 2008 aquaculture accounted for 46% of total food fish supply, around 115 million tonnes. Although wild caught juveniles are still utilised in the industry, concerns over sustainability of extracting juveniles, and the variable timing and magnitude of natural spawning events, make hatchery production an attractive alternative to support the growing demands of aquaculture.

# 3

# Fresh Water Fish Farming

## Fishing Industry in the United States

As with other countries, the 200 nautical miles (370 km) exclusive economic zone (EEZ) off the coast of the United States gives its fishing industry special fishing rights. It covers 11.4 million square kilometres (4.38 million sq mi). This is the largest zone in the world, exceeding the land area of the United States.

According to the FAO, in 2005 the United States harvested 4,888,621 tonnes of fish from wild fisheries and another 471,958 tonnes from aquaculture. This made the United States the fifth leading producer of fish after China, Peru, India and Indonesia, with 3.8 percent of the world total.

### *Management*

Historically, fisheries developed in the U.S. as each area was settled, whether by the original aboriginal peoples or by post Columbian arrivals. Concern for the sustainability of fishery resources was evident as early as 1871, when Congress wrote that "... the most valuable food fishes of the coast and the lakes of the U.S. are rapidly diminishing in number, to the public injury, and so as materially to affect the interests of trade and commerce...." However, it was not until 1976, with the Magnuson-Stevens Fishery Conservation and Management Act, that the federal government began actively managing fisheries.

Today, inland fisheries and nearshore marine fisheries are managed by state (or regional or county) fisheries commissions. State jurisdictions usually extend 3 nautical miles (6 km) out to sea. Coastal fisheries in the EEZ beyond state jurisdictions are the responsibility of the federal system. The primary institutions of the federal system are eight regional fishery management councils and the National Marine Fisheries Service (NMFS), also known as NOAA Fisheries. NMFS works in partnership with industry, universities, and state, local, and tribal agencies to collect data about commercial species.

Fisheries observers on fishing vessels, transmit real time data electronically to NMFS. NMFS works in partnership with the regional fishery management councils to prevent overfishing and restore overfished stocks. Objectives are to reduce fishing intensity, monitor the fisheries, and implement measures to reduce by catch and protect essential fish habitat. NMFS is establishing marine protected areas and individual fishing quotas, and implementing ecosystem based fishery management.

### *Wild Fisheries*

#### *EEZ*

The U.S. EEZ is the largest in the world, 1.7 times the land area of the US, and includes eight large marine ecosystems:

- Northeast U.S. Continental Shelf
- Southeast U.S. Continental Shelf
- Caribbean Sea
- Gulf of Mexico
- California Current
- Insular Pacific-Hawaiian
- Gulf of Alaska.

### *Nearshore Fisheries*

Nearshore fishing areas consist of estuaries and coastal areas that lie within 3 nautical miles (6 km) of the shoreline. These are under the control of the relevant coastal state rather than under federal or NMFS control. They vary widely in species diversity and abundance. Many species are highly prized game fish, while others are small forage fishused for bait, animal food, and industrial products. Those of greatest interest include invertebrate species like crabs, shrimps, abalones, clams, oysters and scallops. Many species are of unknown status because it is difficult to assess their condition. There are no firm estimates exist for long term potential yield.

Management is spread out among the coastal states and other local authorities, and a comprehensive treatment of the fisheries has not been attempted. Traditional techniques are usually employed, including size limits, catch limits, method restrictions, and area and time closures.

### *Coastal Fisheries*

The coastal fisheries are the fisheries that lie within 200 nautical miles (370 km) of the coast that defines the exclusive economic zone, but outside the 3 nautical miles (6 km) distance that defines the nearshore fisheries. Coastal fisheries are under federal control in the form of NMFS, as well as under the control of one of the eight regional fishery management councils

### *Northeast*

Traditionally the most valued fishery for the northeast region has been groundfish. However, the groundfish, especially haddock, yellow tail flounder and cod have been overfished. Record low spawning biomass levels occurred in 1993–94, though these are now recovering. Dogfish and skate rebounded in the 1970s while groundfish and flounder declined. These fish are an important part of the Georges Bank.

The next most important fishery by value is American lobster and Atlantic sea scallop. The striped bass was driven to low levels early in the 1980s. Catch restrictions were applied in the mid 1980s, and by 1995 this an adromous fish had recovered.

The region has valuable mollusk fisheries. Offshore are sea scallops, surfclams, American lobsters and ocean quahog. Inshore are oysters, blue mussels, blue crabs, and clam fisheries. These fisheries are pretty much fully exploited.

The fisheries in the Northeast Region are governed mostly by Fishery Management Plans (FMPs). An example of successful regulation is the 1994 Amendment 4 to the Sea Scallop FMP.

This set out to control scallop fishing effort by not accepting more entrants, by restricting the time vessels could fish each day, and by requiring bigger mesh sizes on dredges. Additional protection for scallops was given closing parts of Georges Bank to groundfish fishing. The scallops recovered.

### Southeast

The Southeast region spans the Gulf of Mexico, the Caribbean Sea and the US Southeast Atlantic. Important species are menhaden, drum, croaker, invertebrates, highly migratory species, reef fish and other nearshore species. Overfishing of king and Spanish mackerel occurred in the 1980s. Regulations were introduced to restrict the size, amount of catch, fishing locations and bag limits for recreational fishers as well as commercial fishers. Gillnets were banned in waters off Florida. By 2001, the mackerel stocks had bounced back. Menhaden and stone crabs and the three major species of shrimp (pink, brown and white) are fully exploited. Spiny lobster are overexploited. Better data is needed if stocks are to be assessed accurately.

### Alaska

The salmon species in Alaska (chinook, coho, pink, sockeye, and chum) generally produce good harvest, though some stocks are declining. Up to 1977, foreign fishing controlled the groundfish fisheries in Alaska (apart from Pacific halibut). The Aleutian Islands, a series of over 300 rocky islands, stretch over 1,000 miles (1,600 km) from southwest Alaska to Russia. They are home to the largest fishing port in the U.S., Dutch Harbour.

The primary target species is pollock, but crabs, salmon and groundfish are also important. Between 1997 and 2001, groundfish were caught bybottom trawling, a fishing method which destroys the habitat of the groundfish, the ocean floor coral and sponge communities.

In 2005, about 80,000 $km^2$ (30,000 sq mi) of seafloor around the Aleutian Islands were permanently closed to destructive fishing gears.

In 2008, NOAA fisheries created the Aleutian Islands Habitat Conservation Area (AIHCA), most of which is closed to bottom trawling. At 725,000 $km^2$ (280,000 sq mi), the AIHCA is the largest marine protected area in the USA.

### Pacific Coast

On the Pacific Coast important species are Pacific halibut, Pacific salmon, groundfish, pelagic fishes and nearshore species. The stocks are mostly overfished or fully exploited.

- Salmon: Salmon production has decreased since the late 1970s, partly due to habitat degradation.

- Groundfish: this harvest is dominated by Pacific whiting. Some stocks need rebuilding. Rockfish can live as long as 100 years, and grow and reproduce very slowly. This makes stock recovery very slow. In 2004, plans for rebuilding several overfished groundfish species were approved .
- Shellfish: Shellfish, such as crabs, clams, shrimp and abalone, sell for high prices, so their fisheries can be small by volume but high by value. These fisheries are mostly fully exploited recreational fishers catch more of some species than the commercial fishers.
- Recovering stocks: After decades of decline, Pacific sardine populations are recovering. In 2004, the NMFS declared that, due to reduced harvest and bycatch, Pacific whiting were recovering.

### Western Pacific

The Western Pacific region includes the western and central Pacific, the Hawaiian Islands, and the islands of the Northern Marianas, Guam and American Samoa. These are tropical or subtropical waters. They have large species diversity but, because ocean nutrients are not rich, the sustainable yields is low. Pelagic armourhead is the only overfished stock.

- Groundfish fisheries: Groundfish, such as snapper jacks, emperors and grouper are harvested from coral and rock habitats, mostly around the Hawaiian Islands. While some stocks are underutilised, other important species have declined to 30 percent of their earlier stock levels.
- Invertebrate fisheries: The main invertebrate fisheries are in the north west Hawaiian Islands for slipper and spiny lobster. This fishery started in 1977, peaked in the mid 1980s, and then declined. The decline is thought to stem from oceanographic changes. Some recovery has occurred since 1991, due to entry and harvest restrictions.
- Highly migratory species account for 99 percent catch in this region and are discussed below.

### High Sea Fisheries

The high seas, or international waters, are the waters outside the jurisdiction of the EEZ of any country. U. S. distant fisheries operate in some of these waters, alongside other nations, beyond the jurisdiction of any nation's coastal fisheries.

### Highly Migratory Species

The main commercial species in the high seas are the highly migratory species. These fish make long migrations across the high seas and are fished by many nations. Highly migratory fish also cross boundaries without regard for international laws. In particular, they enter the EEZ zones of the U.S., which means they become important species also for U.S. coastal fisheries.

In the Atlantic, overfished species include:

- Tuna: bigeye in the Atlantic, albacore in the North Atlantic, and bluefin in the West Atlantic, with yellowfin tuna close to being overfished.
- Blue marlin, white marlin and sailfish.
- Sharks: bull, dusky, bignose, night, Caribbean reef, spinner, silky, tiger, lemon, sand, nurse, scalloped, smooth smooth hammerhead and white shark.

The biomass of swordfish in the North Atlantic has increased, probably due to catch reduction and strong recruitment.

In the Pacific, these transboundary fisheries are also important to other Pacific Rim nations and to U.S. fleets fishing within and beyond the EEZ. The major U.S. catch stock is tuna, though billfish, swordfish and shark are also caught. These stocks account for 99 percent of the Western Pacific region's catch.

### Seamounts

A seamount is a mountain rising from the ocean seafloor that does not reach the ocean surface. They are defined by oceanographers as independent features that rise at least 1,000 metres above the seafloor. Seamounts became interesting during the 1960s when it was discovered that they can maintain large stocks of commercially important fishes and invertebrates.

From 1968 to about 1990, foreign fleets harvested pelagicarmourhead across seamounts in the Hawaiian Ridge. Since 1984, fishing been prohibited there so the stock can recover.

Generally, seamounts are found in the high seas, outside the continental shelves and outside the jurisdiction of any nation's EEZ.

However, the U.S. Pacific Coast region has an unusually small continental shelf, while the Western Coast region contains islands with no continental shelf. As a result, there are some seamounts within U.S. jurisdiction.

### *Inland Fisheries*

Fisheries in inland waters of the United States are small compared to marine fisheries. The largest fisheries are the landings from the Great Lakes, worth about $13 million in 2003, with a similar amount from the Mississippi River basin. This is less than one percent of the dollar value of the marine fisheries.

### *Fishing Fleet*

Until the late 19th century, the U.S. fishing fleet used sailing vessels. By the early 20th century, fishing vessels were built as steam boats with steam engines, or as schooners with auxiliary gasoline engines. By the 1930s the fleet was almost completely converted to diesel vessels.

Fishing gear became more technical: Alaska purse seiners were in use by 1870, longliners were introduced in 1885; otter trawls were operating in the groundfish and shrimp fisheries by the early 20th century. In the late 1960s, factory ships from other countries started fishing haddock, herring, salmon and halibut on traditional U.S. fishing grounds.

Technological advances have played an important role in the development of U.S. fisheries. Increases in size and speed allowed vessels to fish in more distant waters. Advances include double trawls, the Puretic power blocks for retrieving seine nets, refrigerated holds, durable synthetic fibres for lines and nets, GPS to navigate and locate fishing grounds, fishfinders for the location of fish, and spotter planes to locate fish schools.

The vessel size and gear types of the U.S. fishing fleet varies geographically and between fisheries. Because of this, management of U.S. fisheries with a single policy is not feasible. Individual fisheries have their own biological, economic, and sociological characteristics which make broad policies impractical. On the other hand, ad hoc regulations for individual fisheries is not practical either.

U.S. fisheries use most fishing gear types. Vessels are often configured so they can change rapidly between two or more gear types, such as lobster pots to bottom trawls to scallop dredges. The main techniques are purse seining and trawling. Some vessels freeze their catch at sea, such as factory trawlers, tuna boats, Alaskan crab pot vessels, and some southeast shrimp trawlers. Vessels usually land their catches near their homeports.

Based in Hawaii is a longline fishery for tuna and billfish. New management approaches, including placing fisheries observers on vessels to observe what is happening with protected species such as albatross and leatherback, loggerhead and green sea turtles, is expected to reduce incidental catch.

### *Overfished Stocks*

By 2003, overfishing had occurred on 60 stocks. Another 232 stocks were not overfished. Overfishing had been stopped on 31 stocks, and a gain was made of 13 stocks that had been fully rebuilt.

There were 617 other stocks which have limited data or for which criteria for overfishing had not been defined. These stocks mostly have harvests which are not significant, so they are not allocated research funding. Their assessment can be expensive and there is no evidence of overfishing. Rebuilding strategies are operating or are being developed for most stocks which are overfished.

There were 267 major stocks, that is, stocks with at least 200,000 pounds (91,000 kg) in annual landings. Of these, 40 were overfished, 147 were not, and it was not known whether the remaining 80 stocks were overfished. There were 642 minor stocks, of which 20 were overfished, 85 were not, and the remainder had an overfishing status which is unknown or is undefined.

### *Marine Protected Areas*

A marine protected area (MPA) is federally defined as: "any area of the marine environment that has been reserved by federal, state, tribal, territorial, or local laws or regulations to provide lasting protection for part or all of the natural and cultural resources therein."

Fourteen MPAs covering 150,000 square miles (390,000 $km^2$) have been federally-designated as National Marine Sanctuaries, and are administered by a division of the National Oceanic and Atmospheric Administration (NOAA).

In practice, these MPAs are defined areas where natural and/or cultural resources are given more protection than occurs in the surrounding waters. United States MPAs cover many habitats including the open ocean, coastal areas, intertidal zones, estuaries, and the Great Lakes. They vary in purpose, legal authorities, agencies, management approaches, level of protection, and restrictions on human uses.

## Aquaculture

The value of aquaculture products grew from $45 million in 1974 to about $866 million (393,400 tonnes) in 2003. Aquaculture, in the United States, includes the farming of hatchery fish and shellfish which are grown to market size in ponds, tanks, cages, or raceways and released into the wild.

Aquaculture is also used to support commercial and recreational marine fisheries by enhancing or rebuilding wild stock populations. It also includes the cultivation of ornamental fish for the aquarium trade, as well as plant species used in various pharmaceutical, nutritional, and biotechnology products.

According to the FAO, in 2004 the United States ranked 10th in total aquaculture production, behind China, India, Vietnam, Thailand, Indonesia, Bangladesh, Japan, Chile, and Norway. The United States imports aquaculture products from these and other countries, and operates an annual seafood trade deficit of over $9 billion.

Shellfish (oysters, clams, mussels), account for two-thirds of marine aquaculture production, followed by salmon (25 percent) and shrimp (10 percent). Production occurs mainly on land, in ponds, and in coastal waters under state jurisdiction. As a result of recent advances in offshore aquaculture technology, commercial finfish and shellfish operations have been established in more exposed, open ocean locations in state waters in New Hampshire, Hawaii, and Puerto Rico.

Total U.S. aquaculture production, including aquatic plants, is about $1 billion annually, compared to the total world production of about $70 billion. Only about 20 percent of U.S. aquaculture production is from marine species. NOAA estimates that the annual U.S. domestic aquaculture production of all species could increase from about 0.5 million tons to 1.5 million tons by 2025.

## Subsidies

In a study of federal investment in the fishery sector, most aspects of United States tax, fisheries, and societal policies were examined to see whether they created subsidies for the fishing industry and whether these subsidies had impacts. The gross value of direct U.S. subsidies was about 0.5% of the gross ex-vessel value of commercial landings. There are no major ship construction subsidies, market development or other forms of assistance that often exist in developed and developing fishing industries around the world.

### Recreational

In 2006, the U.S. Fish and Wildlife Service estimated that 30.0 million U.S. anglers, 16 years old and older took 403 million fishing trips, spending $42.0 billion in fishing related expenses. Of these, 25.4 million were freshwater anglers who took 337 million trips and spent $26.3 billion. Saltwater fishing attracted 7.7 million anglers who took 67 million trips and spent $8.9 billion.

| ***Recreational fishing expenditures in $US billion*** | | | |
|---|---|---|---|
| | ***subtotal*** | ***total*** | ***percent*** |
| Food and lodging | 6.3 | | |
| Transportation | 5.0 | | |
| Other trip costs: land use fees, guide fees, equipment rental, boating expenses, and bait... | 6.6 | | |
| Trip-related, total | | 17.9 | |
| Fishing equipment: rods, reels, tackle boxes, depth finders, and artificial lures and flies... | 5.3 | | |
| Auxiliary equipment: camping equipment, binoculars, and special fishing clothing... | 0.8 | | |
| Special equipment: boats, vans, and cabins... | 13.2 | | |
| Equipment, total | | 18.8 | |
| Magazines, books | 0.1 | | |
| Membership dues and contributions | 0.2 | | |
| Land leasing and ownership | 4.6 | | |
| Licenses, stamps, tags, and permits | 0.5 | | |
| Other, total | | 5.4 | |
| Overall total | | 42.0 | 100 |

## Magnuson–Stevens Fishery Conservation and Management Act

The Magnuson–Stevens Fishery Conservation and Management Act (MFCMA), commonly referred to as the Magnuson–Stevens Act, is the primary law governing marine fisheries management in the United States. The law is named after Warren G. Magnuson, former U.S. senator from Washington state, and Ted Stevens, the former senator from Alaska.

The Magnuson–Stevens Act was originally enacted as the *Fishery Conservation and Management Act of 1976.* The act has been amended many times over the years. Two major recent sets of amendments to the law were the Sustainable Fisheries Act of 1996 and then 10 years later the *Magnuson–Stevens Fishery Conservation and Management Reauthorisation Act of 2006.*

### Purpose

The MFCMA was enacted to promote the U.S. fishing industry's optimal exploitation of coastal fisheries by "consolidating control over territorial waters" and establishing eight regional councils to manage fish stocks.

The act has been amended several times in response to continued overfishing of major stocks. The most recent version, authorised in 2007, includes seven purposes:

1. Acting to conserve fishery resources
2. Supporting enforcement of international fishing agreements
3. Promoting fishing in line with conservation principles
4. Providing for the implementation of fishery management plans (FMPs) which achieve optimal yield
5. Establishing Regional Fishery Management Councils to steward fishery resources through the preparation, monitoring, and revising of plans which (A) enable stake holders to participate in the administration of fisheries and (B) consider social and economic needs of states.
6. Developing underutilised fisheries
7. Protecting essential fish habitats

Additionally, the law calls for reducing bycatch and establishing fishery information monitoring systems.

### Regulatory Mechanisms

Regional Fishery Management Councils are charged with developing and implementing fishery management plans, both to restore depleted stocks and manage healthy stocks. The National Marine Fisheries Service (NMFS) aides the Secretary of Commerce, who evaluates and approves the council's FMPs. Regional Fishery Management Council members are nominated by the governors of their respective states, and approved by the Secretary of Commerce.

A FMP must specify the criteria which determine when a stock is overfished and the measures needed to rebuild it. Regional councils regulate fishers with mechanisms, including annual catch limits, individual catch limits, community development quotas, and others. The Marine Fish Conservation Network highlighted the most significant changes in the mechanisms utilised in a 2010 report:

> *"To achieve the goal of ending overfishing ... Congress strengthened the role of science in the fishery management process and required fishery managers to establish science based annual catch limits (ACLs) and accountability measures (AMs) for all US fisheries with a deadline of 2010 for all stocks subject to overfishing... The new fisheries law requires the councils' science advisors, the scientific and statistical committees to make recommendations for 'acceptable biological catch' (ABC) which managers may not exceed..."*

The ACL is the centrepiece of the report which is supplemented by other mechanisms regulating the types of gear used, licensing vessels, and using of observers on fishing boats. In section 303 b, the Act enumerates the types of actions authorised for use by councils to achieve optimal catch goals. Including

1. Permitting vessels or operators
2. Designating Zones and periods where fishing is limited
3. Limiting sale, catch or transport of certain fish.
4. Regulating types of fishing equipment
5. Requiring observers onboard vessels.

Regional management councils have taken the controversial step of buying fishing vessels to remove them from the water to reduce the size of the fishing fleet.

### Regulatory Effectiveness

The act's results vary for different regions and different fish stocks. It did not prevent the overfishing of many species throughout its first 20 years of existence. This prompted major amendments in 1996 and 2006. The National Marine Fisheries Service issued a report to Congress in 2010 on the status of U.S. fisheries. It reported that of the 192 stocks monitored for overfishing 38 stocks (20%) still have fish "mortality rates that exceed the overfishing threshold ... and 42 stocks (22%) are overfished". This is down from 38% and 48% respectively in 2000. A 2003 NMFS report reviewed achievements of the act since 1996. Highlighting the inconsistent effects of the legislation, it revealed that overfishing was eliminated in 15 major fish stocks while overfishing was initiated in 12 major fish stocks. To improve their overfishing prevention programs, the NMFS has implemented the Fish Stock Sustainability Index (FSSI), which measures key stocks according to their overfishing status and biomass levels. Since the FSSI began the index has increased every year.

The act also has impacts on financial matters. While taxpayers have paid over US$3 billion on NFMS programs since the bills inception, ultimately the fishing industry is most affected by the act's design flaws and incomplete implementation. According to Zeke Grader, Jr., the executive director of the Pacific Coast Federation of Fishermen's Associations, the largest active trade association of commercial fishermen on the west coast, "Most of the U.S. fisheries stocks are facing a disaster due to over capitalisation of the fishing industry and the mismanagement practices of U.S. Department of Commerce's National Marine Fisheries Service (NMFS) and their appointed regional fishery management councils". Chris Kellogg, chief technical officer for the New England Fishery Management Council emphasizes the tenuous position of many fishers, "On average, my guess is the fishing harvesting industry here in New England is basically covering costs and just on the border of solvency and insolvency".

### Stakeholders

Many interest groups are concerned with the forming of fisheries conservation legislation. Fishers, corporations, activist groups and the public all share interest in protecting fishing eco-systems and economies via the MSFCMA.

Fishers advocate measures that encourage regulatory processes to be scaled to the local level and which ensure fishing privileges aren't concentrated into small groups. They are also aware that if too much competition for finite resources prevails, their livelihood will suffer. Despite this, some fishermen prefer minimal government intervention in their market, defiantly demanding "the right to go broke".

The fishing industry is a $4billion a year industry. Fish harvesting and processing corporations are invested in the political process to maximise their profits, to protect against foreign competition and to prevent regulations from making their proprietary information available to the public. University, argues that federal legislators cannot forge these relationships largely because they haven't properly identified what a "fishing community" is. She highlights the successes of the Community Development Quota system employed in some Alaska fisheries.

Non-Governmental Organisations (NGOs) focusing on fish conservation and environmentalism were key forces behind the incremental improvements in the law's regulatory mechanisms. The Pew Oceans Commission and the US commission on Oceans Policy prompted many of the amendments found within the 2006 reauthorisation. Many advocacy groups speak through coalitions. The Marine Fish Conservation Network, for example, represents over 90 member organisations from across the United States.

The public is represented as a stake holder by elected representatives, who ostensibly take them into consideration when drafting ways to protect public resources such as fish stocks.

### *Critiques*

The major criticisms of the act have been its failure to stem overfishing, minimise bycatch, and to hold accountable regional councils that don't enforce or implement fisheries management plans. Additionally, the act is accused of coddling fishers who make poor business decisions, inadequately protecting ecosystems, and lacking transparency requirements for fishery related data.

Poor enforcement of previous versions of the act is often blamed for depleting stocks. Fishermen have accused each other of cheating on landings and chastised regulators for concentrating the quota allocations into too few hands. Other critics claim the regulatory framework is too "top-down" and alienates local fishers, thereby reducing the likelihood to achieve the cooperation needed to enforce many provisions. Candice May, of Colorado State University, argues that federal legislators can't forge these relationships largely because they haven't properly identified what a "fishing community" is. She highlights the successes of the Community Development Quota system employed in some Alaska fisheries.

In response to these types of criticisms, "On May 1, 2010, the National Marine Fisheries Service (NMFS) implemented a new management system for ground fish in New England. It established 17 fishermen-run collectives, called sectors. Sectors were pioneered by fishermen as voluntary, cooperative and community-based, and were designed to protect fleet diversity and coastal communities. The new management system operates on three simple premises:

- It implements science-based catch limits to rebuild fish populations and prevent overfishing.
- It incorporates monitoring so fishermen and regulators know exactly how much fish is being caught, and as a result, fishing stops once catch limits have been reached.
- Each sector receives its own share of the annual catch. While respecting catch limits, the co-ops provide fishermen with the flexibility to set their own fishing guidelines so they can run their businesses more efficiently and profitably. Those who develop more innovative fishing gear can target more of the healthy fish populations and avoid those populations that are struggling."

As reported at the Council meeting, the first three months of sector operations resulted in (May 1 – August 15):

- Fishermen earning more money for less fishing under the new system. In 2010, landings are down compared to 2009. Only 85.8 percent of total landings last year were landed this year (for the same period of time). Meanwhile, revenues are up 112.4 percent.
- Sector fishermen are avoiding weak stocks and targeting robust stocks. The ratio of Georges Bank cod to Georges Bank haddock (in metric tons) in 2009 was 1121:1532. In 2010, it was 743:2768.
- Landings of Gulf of Maine winter flounder, a stock at very low abundance, are being effectively avoided under sectors. In 2009, 66 metric tons were landed. In 2010, 32 metric tons were landed."

## Fisheries Management

Fisheries management draws on fisheries science in order to find ways to protect fishery resources so sustainable exploitation is possible. Modern fisheries management is often referred to as a governmental system of appropriate management rules based on defined objectives and a mix of management means to implement the rules, which are put in place by a system of monitoring control and surveillance. According to the FAO, there are "no clear and generally accepted definitions of fisheries management". However, the working definition used by the FAO and much quoted elsewhere is:

> *The integrated process of information gathering, analysis, planning, consultation, decision-making, allocation of resources and formulation and implementation, with enforcement as necessary, of regulations or rules which govern fisheries activities in order to ensure the continued productivity of the resources and the accomplishment of other fisheries objectives.*

### History

Fisheries have been explicitly managed in some places for hundreds of years. For example, the Mâori people, New Zealand residents for about 700 years, had prohibitions against taking more than could be eaten and about giving back the first fish caught as an offering to sea god Tangaroa. Starting in the 18th century attempts were made to regulate fishing in the North Norwegian fishery. This resulted in the enactment of a law in 1816 on the Lofoten fishery, which established in some measure what has come to be known as territorial use rights.

> *"The fishing banks were divided into areas belonging to the nearest fishing base on land and further subdivided into fields where the boats were allowed to fish. The allocation of the fishing fields was in the hands of local governing committees, usually headed by the owner of the onshore facilities which the fishermen had to rent for accommodation and for drying the fish."*

Governmental resource protection-based fisheries management is a relatively new idea, first developed for North European fisheries after the first Overfishing Conference held in London in 1936. In 1957 British fisheries researchers Ray Beverton and Sidney Holt published a seminal work on North Sea commercial fisheries dynamics. In the 1960s the work became the theoretical platform for North European management schemes.

After some years away from the field of fisheries management, Beverton criticized his earlier work in a paper given at the first World Fisheries Congress in Athens in 1992. "The Dynamics of Exploited Fish Populations" expressed his concerns, including the way his and

Sydney Holt's work had been misinterpreted and misused by fishery biologists and managers during the previous 30 years. Nevertheless, the institutional foundation for modern fishery management had been laid.

### Political Objectives

According to the FAO, fisheries management should be based explicitly on political objectives, ideally with transparent priorities. Typical political objectives when exploiting a fish resource are to:

- maximise sustainable biomass yield
- maximise sustainable economic yield
- secure and increase employment
- secure protein production and food supplies
- increase export income

Such political goals can also be a weak part of fisheries management, since the objectives can conflict with each other.

### International Objectives

Fisheries objectives need to be expressed in concrete management rules. In most countries fisheries management rules should be based on the internationally agreed, though non-binding, Code of Conduct for Responsible Fisheries, agreed at a meeting of the U.N.'s Food and Agriculture Organisation FAO session in 1995. The precautionary approach it prescribes is typically implemented in concrete management rules as minimum spawning biomass, maximum fishing mortality rates, etc. In 2005 the Fisheries Centre at the University of British Columbia comprehensively reviewed the performance of the world's major fishing nations against the Code.

International agreements are required in order to regulate fisheries in international waters. The desire for agreement on this and other maritime issues led to three conferences on the Law of the Sea, and ultimately to the treaty known as the United Nations Convention on the Law of the Sea (UNCLOS). Concepts such as exclusive economic zones (EEZ, extending 200 nautical miles (370 km) from a nation's coasts) allocate certain sovereign rights and responsibilities for resource management to individual countries.

Other situations need additional intergovernmental coordination. For example, in the Merranean Sea and other relatively narrow bodies of water, EEZ of 200 nautical miles (370 km) are irrelevant. International waters beyond 12-nautical-mile (22 km) from shore require explicit agreements.

Straddling fish stocks, which migrate through more than one EEZ also present challenges. Here sovereign responsibility must be agreed with neighbouring coastal states and fishing entities. Usually this is done through the medium of a regional organisation set up for the purpose of coordinating the management of that stock.

UNCLOS does not prescribe precisely how fisheries confined only to international waters should be managed. Several new fisheries (such as high seas bottom trawling fisheries) are not (yet) subject to international agreement across their entire range. In November 2004 the UN General Assembly issued a resolution on Fisheries that prepared for further development of international fisheries management law.

### Management Mechanisms

Many countries have set up Ministries/Government Departments, named "Ministry of Fisheries" or similar, controlling aspects of fisheries within their exclusive economic zones. Four categories of management means have been devised, regulating either input/investment, or output, and operating either directly or indirectly:

| | *Inputs* | *Outputs* |
|---|---|---|
| Indirect | Vessel licensing | Catching techniques |
| Direct | Limited entry | Catch quota and technical regulation |

Technical means may include:

- prohibiting devices such as bows and arrows, and spears, or firearms
- prohibiting nets
- setting minimum mesh sizes
- limiting the average potential catch of a vessel in the fleet (vessel and crew size, gear, electronic gear and other physical "inputs".
- prohibiting bait
- snagging
- limits on fish traps

- limiting the number of poles or lines per fisherman
- restricting the number of simultaneous fishing vessels
- limiting a vessel's average operational intensity per unit time at sea
- limiting average time at sea.

### Catch Quotas

***Individual Transferable Quota:*** Systems that use *individual transferable quotas* (ITQ), also called individual fishing quota limit the total catch and allocate shares of that quota among the fishers who work that fishery. Fishers can buy/sell/trade shares as they choose. A large scale study in 2008 provided strong evidence that ITQ's can help to prevent fishery collapse and even restore fisheries that appear to be in decline.

### Precautionary Principle

The *Fishery Manager's Guidebook* issued in 2009 by the FAO of the United Nations, advises that the precautionary approach or principle should be applied when "ecosystem resilience and human impact (including reversibility) are difficult to forecast and hard to distinguish from natural changes."

The precautionary principle suggests that when an action risks harm, it should not be proceeded with until it can be scientifically proven to be safe. Historically fishery managers have applied this principle the other way round; fishing activities have not been curtailed until it has been proven that they have already damaged existing ecosystems. In a paper published in 2007, Shertzer and Prager suggested that there can be significant benefits to stock biomass and fishery yield if management is stricter and more prompt.

### Climate Change

In the past, changing climate has affected inland and offshore fisheries and such changes are likely to continue. From a fisheries perspective, the specific driving factors of climate change include rising water temperature, alterations in the hydrologic cycle, changes in nutrient fluxes, and relocation of spawning and nursery habitat. Further, changes in such factors would affect resources at all levels of biological organisation, including the genetic, organism, population, and ecosystem levels.

### Population Dynamics

Population dynamics describes the growth and decline of a given fishery stock over time, as controlled by birth, death and migration. It is the basis for understanding changing fishery patterns and issues such as habitat destruction, predation and optimal harvesting rates. The population dynamics of fisheries has been traditionally used by fisheries scientists to determine sustainable yields.

The basic accounting relation for population dynamics is the BIDE model:

$$N_1 = N_0 + B - D + I - E$$

where $N_1$ is the number of individuals at time 1, $N_0$ is the number of individuals at time 0, $B$ is the number of individuals born, $D$ the number that died, $I$ the number that immigrated, and $E$ the number that emigrated between time 0 and time 1. While immigration and emigration can be present in wild fisheries, they are usually not measured.

Care is needed when applying population dynamics to real world fisheries. In the past, over-simplistic modelling, such as ignoring the size, age and reproductive status of the fish, focusing solely on a single species, ignoring bycatch and physical damage to the ecosystem, has accelerated the collapse of key stocks.

### Elderly Maternal Fish

Traditional management practices aim to reduce the number of old, slow-growing fish, leaving more room and resources for younger, faster-growing fish. Most marine fish produce huge numbers of eggs. The assumption was that younger spawners would produce plenty of viable larvae.

However, 2005 research on rockfish shows that large, elderly females are far more important than younger fish in maintaining productive fisheries. The larvae produced by these older maternal fish grow faster, survive starvation better, and are much more likely to survive than the offspring of younger fish. Failure to account for the role of older fish may help explain recent collapses of some major US West Coast fisheries. Recovery of some stocks is expected to take decades. One way to prevent such collapses is to establish marine reserves, where fishing is not allowed and fish populations age naturally.

### Data Quality

According to fisheries scientist Milo Adkison, the primary limitation in fisheries management decisions is the absence of quality data. Fisheries management decisions are often based on population models, but the models need quality data to be effective. He asserts that scientists and fishery managers would be better served with simpler models and improved data.

The most reliable source for summary statistics is the FAO Fisheries Department.

### Ecopath

Ecopath, with Ecosim (EwE), is an ecosystem modelling software suite. It was initially a NOAA initiative led by Jeffrey Polovina, later primarily developed at the Fisheries Centre of the University of British Columbia. In 2007, it was named as one of the ten biggest scientific breakthroughs in NOAA's 200-year history.

The citation states that Ecopath "revolutionised scientists' ability worldwide to understand complex marine ecosystems". Behind this lies two decades of development work by Villy Christensen, Carl Walters, Daniel Pauly, and other fisheries scientists. As of 2010 there are 6000 registered users in 155 countries. Ecopath is widely used in fisheries management as a tool for modelling and visualising the complex relationships that exist in real world marine ecosystems.

### Human Factors

Managing fisheries is about managing people and businesses, and not about managing fish. Fish populations are managed by regulating the actions of people. If fisheries management is to be successful, then associated human factors, such as the reactions of fishermen, are of key importance, and need to be understood.

Management regulations must also consider the implications for stakeholders. Commercial fishermen rely on catches to provide for their families just as farmers rely on crops.

Commercial fishing can be a traditional trade passed down from generation to generation. Most commercial fishing is based in towns built around the fishing industry; regulation changes can impact an entire town's economy. Cuts in harvest quotas can have adverse effects on the ability of fishermen to compete with the tourism industry.

### Performance

The biomass of global fish stocks has been allowed to run down. This biomass is now diminished to the point where it is no longer possible to sustainably catch the amount of fish that could be caught. According to a 2008 UN report, titled *The Sunken Billions: The Economic Justification for Fisheries Reform*, the world's fishing fleets incur a "$US 50 billion annual economic loss" through depleted stocks and poor fisheries management.

The report, produced jointly by the World Bank and the UN Food and Agriculture Organisation (FAO), asserts that half the world's fishing fleet could be scrapped with no change in catch.

> *"By improving governance of marine fisheries, society could capture a substantial part of this $50 billion annual economic loss. Through comprehensive reform, the fisheries sector could become a basis for economic growth and the creation of alternative livelihoods in many countries. At the same time, a nation's natural capital in the form of fish stocks could be greatly increased and the negative impacts of the fisheries on the marine environment reduced."*

The most prominent failure of fisheries management in recent times has perhaps been the events that lead to the collapse of the northern cod fisheries. More recently, the International Consortium of Investigative Journalists produced a series of journalistic investigations called *Looting the seas*. These detail investigations into the black market for bluefin tuna, the subsidies propping up the Spanish fishing industry, and the overfishing of the Chilean jack mackerel.

### Sustainable Fishery

A conventional idea of a sustainable fishery is that it is one that is harvested at a sustainable rate, where the fish population does not decline over time because of fishing practices.

Sustainability in fisheries combines theoretical disciplines, such as the population dynamics of fisheries, with practical strategies, such as avoiding overfishing through techniques such as individual fishing quotas, curtailing destructive and illegal fishing practices by lobbying for appropriate law and policy, setting up protected areas, restoring collapsed fisheries, incorporating all externalities involved in harvesting marine ecosystems into fishery economics, educating stakeholders and the wider public, and developing independent certification programs.

Some primary concerns around sustainability are that heavy fishing pressures, such as overexploitation and growth or recruitment overfishing, will result in the loss of significant potential yield; that stock structure will erode to the point where it loses diversity and resilience to environmental fluctuations; that ecosystems and their economic infrastructures will cycle between collapse and recovery; with each cycle less productive than its predecessor; and that changes will occur in the trophic balance (fishing down marine food webs).

> *"Sustainable management of fisheries cannot be achieved without an acceptance that the long-term goals of fisheries management are the same as those of environmental conservation" —Daniel Pauly and Dave Preikshot,*

Global wild fisheries are believed to have peaked and begun a decline, with valuable habitats, such as estuaries and coral reefs, in critical condition. Current aquaculture or farming of piscivorous fish, such as salmon, does not solve the problem because farmed piscivores are fed products from wild fish, such as forage fish. Salmon farming also has major negative impacts on wild salmon. Fish that occupy the highertrophic levels are less efficient sources of food energy.

Fishery ecosystems are an important subset of the wider marine environment. This article documents the views of fisheries scientists and marine conservationists about innovative approaches towards sustainable fisheries.

### History

> *"In the end, we will conserve only what we love; we will love only what we understand; and we will understand only what we are taught" —Senegalese conservationist Baba Dioum,*

In his 1883 inaugural address to the International Fisheries Exhibition in London, Thomas Huxley asserted that overfishing or "permanent exhaustion" was scientifically impossible, and stated that probably "all the great sea fisheries are inexhaustible". In reality, by

1883 marine fisheries were already collapsing. The United States Fish Commission was established 12 years earlier for the purpose of finding why fisheries in New England were declining. At the time of Huxley's address, the Atlantic halibut fishery had already collapsed (and has never recovered).

### *Traditional Management of Fisheries*

Traditionally, fisheries management and the science underpinning it was distorted by its "narrow focus on target populations and the corresponding failure to account for ecosystem effects leading to declines of species abundance and diversity" and by perceiving the fishing industry as "the sole legitimate user, in effect the owner, of marine living resources."

Historically, stock assessment scientists usually worked in government laboratories and considered their work to be providing services to the fishing industry. These scientists dismissed conservation issues and distanced themselves from the scientists and the science that raised the issues. This happened even as commercial fish stocks deteriorated, and even though many governments were signatories to binding conservation agreements.

### *Defining Sustainability*

The notion of sustainable development is sometimes regarded as an unattainable, even illogical notion because development inevitably depletes and degrades the environment. Ray Hilborn, of the University of Washington, distinguishes three ways of defining a sustainable fishery.

- *Long term constant yield* is the idea that undisturbed nature establishes a steady state that changes little over time. Properly done, fishing at up to maximum sustainable yield allows nature to adjust to a new steady state, without compromising future harvests. However, this view is naive, because constancy is not an attribute of marine ecosystems, which dooms this approach. Stock abundance fluctuates naturally, changing the potential yield over short and long term periods.
- *Preserving intergenerational equity* acknowledges natural fluctuations and regards as unsustainable only practices which damage the genetic structure destroy habitat, or deplete stock levels to the point where rebuilding requires more than a

single generation. Providing rebuilding takes only one generation, overfishing may be economically foolish, but it is not unsustainable. This definition is widely accepted.

- *Maintaining a biological, social and economic system* considers the health of the human ecosystem as well as the marine ecosystem. A fishery which rotates among multiple species can deplete individual stocks and still be sustainable so long as the ecosystem retains its intrinsic integrity. Such a definition might consider as sustainable fishing practices that lead to the reduction and possible extinction of some species.

### Social Sustainability

Fisheries and aquaculture are, directly or indirectly, a source of livelihood for over 500 million people, mostly in developing countries. While biodiversity is important, people need food security.

> *Social sustainability can conflict with biodiversity. A fishery is socially sustainable if the fishery ecosystem maintains the ability to deliver products the society can use. Major species shifts within the ecosystem could be acceptable as long as the flow of such products continues. Humans have been operating such regimes for thousands of years, transforming many ecosystems, depleting or driving to extinction many species. —Ray Hilborn,*

According to Hilborn, the "loss of some species, and indeed transformation of the ecosystem is not incompatible with sustainable harvests."

For example, in recent years, barndoor skates have been caught as bycatch in the western Atlantic. Their numbers have severely declined and they will probably go extinct if these catch rates continue. Even if the barndoor skate goes extinct, changing the ecosystem, there could still be sustainable fishing of other commercial species.

### Reconciling Fisheries with Conservation

At the Fourth World Fisheries Congress in 2004, Daniel Pauly asked, "How can fisheries science and conservation biology achieve a reconciliation?", then answered his own question, "By accepting each other's essentials: that fishing should remain a viable occupation; and that aquatic ecosystems and their biodiversity are allowed to persist."

## Obstacles

### Overfishing

Overfishing can be sustainable. According to Hilborn, overfishing can be "a misallocation of societies' resources", but it does not necessarily threaten conservation or sustainability".

Overfishing is traditionally defined as harvesting so many fish that the yield is less than it would be if fishing were reduced. For example, Pacific salmonare usually managed by trying to determine how many spawning salmon, called the "escapement", are needed each generation to produce the maximum harvestable surplus. The optimum escapement is that needed to reach that surplus. If the escapement is half the optimum, then normal fishing looks like overfishing. But this is still sustainable fishing, which could continue indefinitely at its reduced stock numbers and yield. There is a wide range of escapement sizes that present no threat that the stock might collapse or that the stock structure might erode.

On the other hand, overfishing can precede severe stock depletion and fishery collapse. Hilborn points out that continuing to exert fishing pressure while production decreases, stock collapses and the fishery fails, is largely "the product of institutional failure."

Today over 70% of fish species are either fully exploited, overexploited, depleted, or recovering from depletion. If overfishing does not decrease, it is predicted that stocks of all species currently commercially fished for will collapse by 2048."

A Hubbert linearisation (Hubbert curve) has been applied to the whaling industry, as well as charting the price of caviar, which depends on sturgeon stocks. Another example is North Sea cod. Comparing fisheries and mineral extraction tells us that human pressure on the environment is causing a wide range of resources to go through a Hubbert depletion cycle.

### Habitat Modification

Nearly all the world's continental shelves, and large areas of continental slopes, underwater ridges, and seamounts, have had heavy bottom trawls and dredges repeatedly dragged over their surfaces. For fifty years, governments and organisations, such as the Asian Development Bank, have encouraged the fishing industry to develop trawler fleets. Repeated bottom trawling and dredging literally flattens diversity in the benthic habitat, radically changing the associated communities.

### Changing the Ecosystem Balance

Since 1950, 90 percent of 25 species of big predator fish have gone.

- How we are emptying our seas *The Sunday Times*, May 10, 2009.
- Pauly, Daniel (2004) Reconciling Fisheries with Conservation: the Challenge of Managing Aquatic Ecosystems Fourth World Fisheries Congress, Vancouver, 2004.

### Climate Change

Rising ocean temperatures and ocean acidification are radically altering aquatic ecosystems. Climate change is modifying fish distribution and the productivity of marine and freshwater species.

This reduces sustainable catch levels across many habitats, puts pressure on resources needed for aquaculture, on the communities that depend on fisheries, and on the oceans' ability to capture and store carbon (biological pump). Sea level rise puts coastal fishing communities at risk, while changing rainfall patterns and water use impact on inland (freshwater) fisheries and aquaculture.

### Ocean Pollution

A recent survey of global ocean health concluded that all parts of the ocean have been impacted by human development and that 41 percent has been fouled with human polluted runoff, overfishing, and other abuses. Pollution is not easy to fix, because pollution sources are so dispersed, and are built into the economic systems we depend on.

The United Nations Environment Programme (UNEP) mapped the impacts of stressors such as climate change, pollution, exotic species, and over-exploitation of resources on the oceans. The report shows at least 75 percent of the world's key fishing grounds may be affected.

### Diseases and Toxins

Large predator fish contain significant amounts of mercury, a neurotoxin which can affect fetal development, memory, mental focus, and produce tremors.

### *Irrigation*

Lakes are dependent on the inflow of water from its drainage basin. In some areas, aggressive irrigation has caused this inflow to decrease significantly, causing water depletion and a shrinking of the lake. The most notable example is the Aral Sea, formerly among the four largest lakes in the world, now only a tenth of its former surface area.

## Remediation

### *Fisheries Management*

Fisheries management draws on fisheries science to enable sustainable exploitation. Modern fisheries management is often defined as mandatory rules based on concrete objectives and a mix of management techniques, enforced by a monitoring control and surveillance system.

- *Ideas and rules:* Economist Paul Romer believes sustainable growth is possible providing the right ideas (technology) are combined with the right rules, rather than simply hectoring fishers. There has been no lack of innovative ideas about how to harvest fish. He characterises failures as primarily failures to apply appropriate rules.
- *Fishing subsidies:* Government subsidies influence many of the world fisheries. Operating cost subsidies allow European and Asian fishing fleets to fish in distant waters, such as West Africa. Many experts reject fishing subsidies and advocate restructuring incentives globally to help struggling fisheries recover.
- *Economics:* Another focus of conservationists is on curtailing detrimental human activities by improving fisheries' market structure with techniques such as salable fishing quotas, like those set up by the Northwest Atlantic Fisheries Organisation, or laws such as those listed below.
- *Sustainable fisheries certification:* A promising direction is the independent certification programs for sustainable fisheries conducted by organisations such as the Marine Stewardship Council and Friend of the Sea. These programs work at raising consumer awareness and insight into the nature of their seafood purchases.
- *Ecosystem based fisheries:*

## Ecosystem Based Fisheries

> *"We propose that rebuilding ecosystems, and not sustainability per se, should be the goal of fishery management. Sustainability is a deceptive goal because human harvesting of fish leads to a progressive simplification of ecosystems in favour of smaller, high turnover, lower trophic level fish species that are adapted to withstand disturbance and habitat degradation." — Tony Pitcher and Daniel Pauly,*

According to marine ecologist Chris Frid, the fishing industry points to marine pollution and global warming as the causes of recent, unprecedented declines in fish populations. Frid counters that overfishing has also altered the way the ecosystem works. "Everybody would like to see the rebuilding of fish stocks and this can only be achieved if we understand all of the influences, human and natural, on fish dynamics." He adds: "fish communities can be altered in a number of ways, for example they can decrease if particular-sized individuals of a species are targeted, as this affects predator and prey dynamics. Fishing, however, is not the sole cause of changes to marine life—pollution is another example....No one factor operates in isolation and components of the ecosystem respond differently to each individual factor."

The traditional approach to fisheries science and management has been to focus on a single species. This can be contrasted with theecosystem-based approach. Ecosystem-based fishery concepts have been implemented in some regions. In a 2007 effort to "stimulate much needed discussion" and "clarify the essential components" of ecosystem-based fisheries science, a group of scientists offered the following ten commandments for ecosystem-based fisheries scientists:

- Keep a perspective that is holistic, risk-adverse and adaptive.
- Maintain an "old growth" structure in fish populations, since big, old and fat female fish have been shown to be the best spawners, but are also susceptible to overfishing.
- Characterise and maintain the natural spatial structure of fish stocks, so that management boundaries match natural boundaries in the sea.
- Monitor and maintain seafloor habitats to make sure fish have food and shelter.

- Maintain resilient ecosystems that are able to withstand occasional shocks.
- Identify and maintain critical food-web connections, including predators and forage species.
- Adapt to ecosystem changes through time, both short-term and on longer cycles of decades or centuries, including global climate change.
- Account for evolutionary changes caused by fishing, which tends to remove large, older fish.
- Include the actions of humans and their social and economic systems in all ecological equations.

### *Marine Protected Areas*

Strategies and techniques for marine conservation tend to combine theoretical disciplines, such as population biology, with practical conservation strategies, such as setting up protected areas, as with Marine Protected Areas (MPAs) or Voluntary Marine Conservation Areas. Each nation defines MPAs independently, but they commonly involve increased protection for the area from fishing and other threats. Marine life is not evenly distributed in the oceans.

Most of the really valuable ecosystems are in relatively shallow coastal waters, above or near the continental shelf, where the sunlit waters are often nutrient rich fromland runoff or upwellings at the continental edge, allowing photosynthesis, which energises the lowest trophic levels. In the 1970s, for reasons more to do with oil drilling than with fishing, the U.S. extended its jurisdiction, then 12 miles from the coast, to 200 miles. This made huge shelf areas part of its territory. Other nations followed, extending national control to what became known as the exclusive economic zone (EEZ). This move has had many implications for fisheries conservation, since it means that most of the most productive maritime ecosystems are now under national jurisdictions, opening possibilities for protecting these ecosystems by passing appropriate laws.

Daniel Pauly characterises marine protected areas as "a conservation tool of revolutionary importance that is being incorporated into the fisheries mainstream." The Pew Charitable Trusts have funded various initiatives aimed at encouraging the development of MPAs and other ocean conservation measures.

### Fish Farming

To what extent can farmed fish be part of the answer? Farmed salmon eat three pounds of wild fish to produce one pound of salmon.

### Laws and Treaties

International laws and treaties related to marine conservation include the 1966 Convention on Fishing and Conservation of Living Resources of the High Seas. United States laws related to marine conservation include the 1972 Marine Mammal Protection Act, as well as the 1972 Marine Protection, Research and Sanctuaries Act which established the National Marine Sanctuaries program. Magnuson-Stevens Fishery Conservation and Management Act.

### Awareness Campaigns

Various organisations promote sustainable fishing strategies, educate the public and stakeholders, and lobby for conservation law and policy.

The list includes the Marine Conservation Biology Institute and Blue Frontier Campaign in the U.S., The U.K.'s Frontier (the Society for Environmental Exploration) and Marine Conservation Society, Australian Marine Conservation Society, International Council for the Exploration of the Sea(ICES), Langkawi Declaration, Oceana, PROFISH, and the Sea Around Us Project, International Collective in Support of Fishworkers, World Forum of Fish Harvesters and Fish Workers, Frozen at Sea Fillets Association and CEDO.

The United Nations Millennium Development Goals include, as goal #7: target 2, the intention to "reduce biodiversity loss, achieving, by 2010, a significant reduction in the rate of loss", including improving fisheries management to reduce depletion of fish stocks.

Some organisations certify fishing industry players for sustainable or good practices, such as the Marine Stewardship Council and Friend of the Sea.

Other organisations offer advice to members of the public who eat with an eye to sustainability. According to the marine conservation biologist Callum Roberts, four criteria apply when choosing seafood:

- Is the species in trouble in the wild where the animals were caught?
- Does fishing for the species damage ocean habitats?

- Is there a large amount of bycatch taken with the target species?
- Does the fishery have a problem with discards—generally, undersized animals caught and thrown away because their market value is low?

The following organisations have download links for wallet-sized cards, listing good and bad choices:

- Monterey Bay Aquarium Seafood Watch, USA
- Blue Ocean Institute, USA
- Marine Conservation Society, UK
- Australian Marine Conservation Society
- The Southern African Sustainable Seafood Initiative.

## Data Issues

### *Data Quality*

One of the major impediments to the rational control of marine resources is inadequate data. According to fisheries scientist Milo Adkison (2007), the primary limitation in fisheries management decisions is poor data. Fisheries management decisions are often based on population models, but the models need quality data to be accurate. Scientists and fishery managers would be better served with simpler models and improved data.

### *Unreported Fishing*

Estimates of illegal catch losses range between $10bn and $23bn annually, representing between 11 and 26 million tonnes.

- Incidental catch.

### *Shifting Baselines*

Shifting baselines is a term which describes the way significant changes to a system are measured against previous baselines, which themselves may represent significant changes from the original state of the system. The term was first used by the fisheries scientist Daniel Pauly in his paper "Anecdotes and the shifting baseline syndrome of fisheries". Pauly developed the term in reference to fisheries management where fisheries scientists sometimes fail to identify the correct "baseline" population size (e.g. how abundant a fish species population was *before* human exploitation) and thus work with a

**shifted baseline**. He describes the way that radically depleted fisheries were evaluated by experts who used the state of the fishery at the start of their careers as the baseline, rather than the fishery in its untouched state. Areas that swarmed with a particular species hundreds of years ago, may have experienced long term decline, but it is the level of decades previously that is considered the appropriate reference point for current populations.

In this way large declines in ecosystems or species over long periods of time were, and are, masked. There is a loss of perception of change that occurs when each generation redefines what is "natural".

### *Looting the Seas*

*Looting the seas* is the name given by the International Consortium of Investigative Journalists to a series of journalistic investigations into areas directly affecting the sustainability of fisheries. So far they have investigated three areas involving fraud, negligence and overfishing:

- The black market in bluefin tuna
- Subsidies propping up the Spanish fishing industry
- Overfishing of the southern jack mackerel.

## Aquaculture in South Africa

Aquaculture is about the emerging aquaculture industry in South Africa.

### *Definition of Aquaculture*

Aquaculture is defined as the "controlled cultivation (farming/ production) of a variety of fish, crustaceans, molluscs, and aquatic plants in marine and freshwater mainly for human consumption, industrial use and recreational purposes. In South Africa it consists mainly of freshwater species such as crocodiles, trout, catfish, tilapia and ornamental fish as well as marine species such as abalone, prawns, oysters and mussels."

### *Overview*

According to the Department of Agriculture, Forestry and Fisheries, "Aquaculture in South Africa is divided into freshwater aquaculture and marine aquaculture. Freshwater fish culture is severely limited by the supply of suitable water. The most important areas for the production of fresh water species are the Limpopo, Mpumalanga Lowveld and Northern Kwazulu–Natal. Trout is farmed along the high mountain in Lydenburg area, Kwazulu-Natal Drakensberg and the Western Cape. Other freshwater species cultivated on a small scale include catfish, freshwater crayfish and tilapia species. Marine aquaculture is a fast developing sector, with a focus on mussels, oysters, abalone, seaweeds and prawns. Of these, mussel farming is the best established. Abalone culture is now well established, centred in the Hermanus area on the Cape south coast. There is also an experimental offshore farm (cage culture) off Gansbaai for salmon."

Export data from the Department of Agriculture, Forestry and Fisheries indicate that the South African industry is dominated by the Western Cape province, which accounts for more than 80% of all South African aquaculture produce, followed by the Eastern Cape at a distant 12.75%.

The Department of Trade and Industry states in its sector diagnostic study that national aquaculture production data (2003–2006) in 2003 was 3,485 tons and in 2006 was 3,564 tons which had a value of R210. Abalone accounted for the biggest increase with production increasing by 61% from 515 tons in 2003, to 833 tons in 2006. Declines in production were experienced by the following sub-sectors: oyster, mussel and trout with reductions in production of 19.2%, 39.5% and 18.4% respectively.

### *Position of the South African Industry in Relation to the Global Industry*

The industry in South Africa is in its infancy, with production volumes being very low, even compared to its continental peers where Egypt and Nigeria lead the pack.

### *Government Policy and Support Towards the Sector*

Aquaculture has been identified as a critical industry, due to the popularity of its produce and the declining yields world-wide. In South Africa, Aquaculture has been identified as part of the key industries for promotion in line with the country's Industrial Policy Action Plan II (IPAP II). The Department of Agriculture, Forestry and Fisheries is also lending its support to the sector and encouraging collaboration with stakeholders across the sector.

## Aquaculture in New Zealand

Aquaculture started in New Zealand in the late 1960s. It is dominated by mussels, oysters and salmon.

In 2007, aquaculture generated about NZ$360 million in sales on an area of 7,700 hectares. $240 million was earned in exports.

In 2006, the aquaculture industry in New Zealand developed a strategy aimed at achieving a sustainable annual billion NZ dollar business by 2025. In 2007, the government reacted by offering more support to the growing industry.

Aquaculture is the general term given to the cultivation of any fresh or salt water plant or animal. It takes place in New Zealand in coastal marine areas (mariculture) and in inland tanks or enclosures.

Aquaculture in New Zealand currently (2008) occupies 14,188 ha. Of that area, 7,713 ha is in established growing areas and is owned by the aquaculture industry, 4,010 ha is used to enhance the wild scallop fishery and belongs to the Challenger Scallop Enhancement Company, and 2,465 ha is an exposed site six kilometres offshore from Napier where trials are being undertaken by a private company to test the site's economic viability.

In 2005 the aquaculture industry provided direct employment for about 2,500 full time equivalents, mostly in the processing sector. A similar amount of indirect employment resulted from flow-on effects. The aquaculture industry is important for some coastal areas around New Zealand where there is limited employment. This applies particularly to some Mâoricommunities with traditional links to coastal settlements.

Marine aquaculture, mariculture, occurs in the sea, generally in sheltered bays along the coast. In New Zealand, about 70 percent of marine aquaculture occurs in the top of the South Island. In the North Island, the Firth of Thames is productive.

### *Cultured Species*

There are three main species in the New Zealand aquaculture industry: the green-lipped mussel, the Pacific oyster and king salmon. In 2006 these three species generated $357 million in sales. Mussel accounted for 63 percent of this value, Pacific oysters 9 percent and king salmon 28 percent.

Over two-thirds of New Zealand's aquaculture product comes from mussels and oysters. These shellfish are cultivated in two distinct stages; first spat needs to be collected, then the spat is grown in a grow-out facility.

- Spat, also called seed, is the free-swimming larval stage of a shellfish. Spat is cultured in hatcheries, and can be grown in tanks on land. Hatcheries can also be associated with research facilities where spat can be selectively bred to specifications, as broodstock.
- A grow-out facility is the place where the spat are raised to market size, usually in enclosures anchored in coastal waters.

### *Mussels*

Until the early 1960s, mussels were harvested by hand from intertidal rocks. Dredging was then introduced, and within a few years the mussel beds in Tasman Bay and the Hauraki Gulf were dredged clean.

In the late 1960s, following this collapse, the aquaculture of the New Zealand mussel began. The endemic green lipped mussel was used to trial growing mussel spat (young mussels) on ropes suspended from rafts.

The Hauraki Gulf and the Marlborough Sounds provided sheltered environments, with clean water rich in plankton. The cultured mussels were ready for harvest after 12 to 18 months, and first went on sale in 1971.

More growers entered the industry. The labour-intensive raft method was replaced with a modified Japanese longline system. Biodegradable stockings were packed with spat and tied to parallel rows of looped ropes, supported by buoys. Young mussels grow through the stockings, anchoring themselves to the ropes with their strong byssal threads (beards).

The farms are usually located in sheltered or semi-sheltered areas where there is sufficient depth of water at low tide to keep the longline droppers off the bottom. Recent research has been investigating offshore mussel farming in exposed areas several kilometres from shore, such as farms offshore from Napier and Opotiki.

Initially the ropes were allowed to reseed naturally, after harvest, from the spat already present in coastal waters. However, this method was unreliable. In 1974 a marine scientist discovered mussel spat encrusted on drift kelp on Ninety Mile Beach. Locals collected the seaweed and air freighted it to mussel farmers. Kaitaia spat, as it became known, is now the prime source of seed mussels. There are some experimental hatcheries.Improved techniques have led to rapid production increases, and bulk handling methods have been introduced to meet growing demand for export to more than 60 countries. By 2006 there were over 900 mussel farms in New Zealand covering about 6500 hectares, and worth about $224 million in annual sales. About $180 million were exports, usually sold as frozen mussels in half shells, patented with the trade name NZ Greenshell Mussels.

***Figure:** Green lipped mussel*

### Oysters

There are two types of wild oysters in New Zealand, Bluff oysters and rock oysters. Both have been commercially harvested since the mid-19th century. Bluff oysters have never been cultivated, but various attempts were made to cultivate the rock oyster.

Rock oysters are found naturally in the intertidal zone in the north of the North Island, and were subject to early cultivation experiments. During the 1960s, commercial farmers grew rock oysters on sticks coated with cement, and laid in racks in the lower intertidal regions of harbours and inlets around the northern North Island.

Then in 1970 another oyster started outgrowing the native rock oyster. This newcomer was the Pacific oyster, which had probably been introduced into New Zealand waters in the 1950s from a Japanese vessel hull or in their ballast water. At first, farmers tried to remove it from their collecting sticks, but year by year Pacific oyster spat increased, out-competing the native rock oyster.

Eventually commercial growers began to cultivate the Pacific oyster, and by the mid-1970s, it had become the main farm-raised oyster. Pacific oysters have well-established international markets, grow three times faster than native rock oysters, reach a larger size, have several spawnings each year and produce more consistent quantities of spat.

In 1977 Pacific oysters appeared in the Marlborough Sounds, and farming began there in the 1990s. Instead of using the North Island method of cultivating oysters on racks, Marlborough farmers used hanging longlines, a system developed in Australia.

By 2006 there were over 230 oyster farms in New Zealand using over 900 hectares, and worth about $32 million in annual sales. About $18 million were exports.

Hatchery reared seed is not suitable for the rack systems which are still used by much of the industry. Pacific oysters on these racks are cultured from wild spat, most of which is gathered in the Kaipara Harbour. However, other systems are increasingly being used, including hanging longlines and plastic trays. The Cawthron Institute is the main provider of hatchery spat, and can selectively breed spat to specifications.

### Salmon

Around 1900, different salmon were introduced as sport fish. Only the king salmon (Chinook) adapted to the environment. For decades, the development of salmon and trout aquaculture in New Zealand was opposed by recreational fishers, on the grounds that disease would spread from fish farms into recreational fisheries, and that wild fish would be poached if they could be sold. In 1973 the government compromised by making trout farms illegal, but salmon farms legal. New Zealand is probably the only country in the world where trout farming is illegal, despite commercial interest from companies already farming salmon.

In 1976, the first salmon farm was established at Pupu Springs, Tasman. Salmon were raised in fresh water, growing to 25 centimetres over two years. The venture was originally aimed at ocean ranching, where juvenile salmon would be released into the sea with the expectation that some would return as adults. But few did return, so the Pupu Springs facility was converted to a hatchery, supplying stock to sea farms.

In 1983, the first sea-cage salmon farm was established in Big Glory Bay, Stewart Island. It was followed by farms in the Marlborough Sounds and at Akaroa, Banks Peninsula. These areas accounted for over 90 percent of the 8,500 tonnes of salmon produced in 2001. Today, New Zealand accounts for over half of the world production of king salmon (7,400 tonnes in 2005).

Farming in the sea (mariculture) for king salmon is sometimes called sea-cage ranching. Sea-cage ranching takes place in large floating net cages, about 25 metres across and 15 metres deep, moored to the sea floor in clean, fast-flowing coastal waters. Smolt (young fish) from freshwater hatcheries are transferred to cages containing several thousand salmon, and remain there for the rest of their life. They are fed fishmeal pellets high in protein and oil. Most of this fishmeal is imported from Australia. The salmon are harvested when they are about two years old, weighing 2.5 to 4 kilograms. Sea cages are located in the Marlborough Sounds, Akaroa Harbour and Stewart Island.

Farming in freshwater for king salmon uses net cages placed in rivers, using techniques similar to those used for sea-farmed salmon. Freshwater raceways are located in several Canterbury rivers such as the Cluthaand Waimakariri Rivers. In the 1990s, a unique form of freshwater salmon farming was developed in hydroelectric canals in the Mackenzie Basin. Young salmon are enclosed in pens in the Ohau and Tekapo canals. The Tekapo site, fed by fast cold waters from the Southern Alps, is the highest salmon farm in the world, 677 metres above sea level.

Before they are killed, cage salmon are anaesthetised with a herbal extract. They are then spiked in the brain. The heart beats for a time as the animal is bled from its sliced gills. Relaxing the salmon like this when it is killed produces firm, long-keeping flesh.

Lack of disease in wild populations and low stocking densities used in the cages means that New Zealand salmon farmers do not use antibiotics and chemicals that are often needed elsewhere.

The New Zealand industry has grown into the largest producers of farmed king salmon in the world, accounting for about half of world production.

The New Zealand King Salmon Company, dominates the production of king salmon in New Zealand. The company has its own selective breeding programmes integrated with quality control and production. Other salmon producers rely on stock from hatcheries where selective breeding is less well developed.

### *Other Species*

- The culture of ornamental cold water species, such as goldfish, are valued at about $18 million.
- A small-scale freshwater prawn farm was established in 1991 at Wairakei, near Taupo, producing tropical giant river prawns. Heat from a geothermal source is used to heat water in prawn-rearing ponds.

Other species which have potential, but are currently small-scale or are still in research or pre-commercial stages, are discussed under Prospects below.

### *Scallop Enhancement*

Enhancement is the name given to techniques designed to boost the natural recruitment or survival of young animals or seaweed in the wild. In New Zealand, scallop enhancement has worked well in Tasman Bay and Golden Bay.

The New Zealand scallop is a large fan-shaped shellfish, flat on one side and convex on the other. It lives on the bottom of coastal waters, 30 metres or more deep. Scallop spat-collecting bags are suspended during summer in coastal areas with high natural scallop settlement. The scallop larvae settle out of the plankton onto the fine feathery surface of the plastic mesh bags.

The larvae are allowed to grow to a suitable size and are then released onto known natural scallop beds at densities of about six per square metre of sea floor. There, they are later harvested on a rotational basis by dredges. This technique has resulted in a marked increase and stabilising of the available annual catch. The Tasman scallop fishery, near collapse in the 1980s, recovered with re-seeding to a level where 747 tonnes were harvested in 2004.

### Legislation and Administration

Marine farmers usually look for sheltered and unpolluted waters rich in nutrients. Often these areas are also desirable for other purposes. In the late 1990s, demand for coastal aquaculture space upsurged, increasing fivefold. Aquaculture consents developed haphazardly, with regional councils unsure about how marine farms might impact coastal environments. By 2001, some councils were inundated with marine farm applications, and were operating with inadequate guidelines for sustainably managing the coast. As the Ministry for the Environment put it: "Attempts to minimise local or cumulative environmental effects resulted in bottlenecks, delays and high costs in processing applications for new marine farms, local moratoria, submitter fatigue and poor environmental outcomes. Marine farmers, local communities, and the government wanted change."

In 2002, the government stopped issuing consents for more new marine farms while they reformed the legislation. The consents had operated under a system overseen by both the Ministry of Fisheries and the regional councils. The reforms aimed to streamline these applications for both freshwater and marine farms. Industry farmers objected to the moratorium, on the grounds that delaying expansion and diversification could not be in the interest of the industry. Mâori groups considered they were especially affected since they were the main applicants for coastal farms.

This took three years, and in early 2005, Parliament passed the Aquaculture Reform Act 2004, which introduced the new legislation. The act amends five existing acts to cope with the new environmental demands, and creates two new acts, the Mâori Commercial Aquaculture Claims Settlement Act 2004 and the Aquaculture Reform (Repeals and Transitional Provisions) Act 2004. The legislation and administration of aquaculture in New Zealand is complex for such a small industry. A more comprehensive overview can be found here.

Aquaculture is administered in New Zealand through labyrinth bureaucracies, with consequent diluted responsibilities. No single ministerial portfolio or government agency is responsible. As an example, in 2007 the government released a strategy on aquaculture. This strategy was endorsed by six government ministers with the following portfolios: fisheries, environment, conservation, local government, Mâori affairs, industry and regional development. Further, there were five government departments directly involved in the preparation of the strategy. As another example, the access to marine and freshwater aquaculture sites are under the control of 17 regional local government agencies with yet more oversight by various central government agencies.

Despite many further consultations and incentives, no new aquaculture space was created under the new legislation for another four years. This coincided with a change in government at the end of 2008, which announced that the aquaculture reforms are to be overhauled.

### *Training and Research*

In recent years, skill levels in the New Zealand aquaculture industry has considerably improved. This has been largely due to Seafood Industry Training Organisation (SITO), an integral part of the seafood industry. SITO have developed tailored aquaculture training programmes based on their prior experience with industry-based training for wild fisheries. They now offer nationally recognised training programmes based on the needs of companies involved in aquaculture.

At the tertiary level, the Auckland University of Technology offers an undergraduate degree in aquaculture. Other tertiary training centres offering aquaculture courses include the Bay of Plenty Polytechnic, the Nelson Marlborough Institute of Technology, and the Mahurangi Technical Institute.

Government funding for aquaculture research is about two percent of the annual sales of the industry. These funds are mostly delivered through a competitive bidding process, organised and controlled by the Foundation for Research, Science and Technology.

- The principal aquaculture research group is the National Institute of Water and Atmospheric Research (NIWA). NIWA is structured as a profit-making private company, though it is owned by the government. It operates three aquaculture research facilities; Bream Bay Aquaculture Park, Mahanga Bay Aquaculture Research Facility and Silverstream Hatchery. The Bream Bay Aquaculture Park includes other private aquaculture companies organised as an industrial-technology park. NIWA produces yellowtail kingfish spat, seed abalone and salmon smolts, which it sells to on-growers.

- The Cawthron Institute is a non-profit organisation which does regional research around Nelson. It operates a nearby saltwater research facility called the Cawthron Aquaculture Park.
- The tertiary education sector undertakes a small amount of aquaculture research. In 2007 an aquaculture centre was opened at Mahurangi Technical Institute in Warkworth. Scientists at the institute are aiming to breed short-fin eels within two years with a goal of producing commercial quantities of eels in captivity, which would be a world first.

### Role of Mâori

In pre-European times, the indigenous Mâori of New Zealand undertook rudimentary aquaculture activities, such as placing suitable rocks into the intertidal settlement zones of oyster larvae. They were also thought to have transplanted abalone and other shellfish between different areas.

Mâori currently have a significant presence in the New Zealand aquaculture industry, and this is likely to increase over time as the requirements to allocate aquaculture space through the Mâori Commercial Aquaculture Claims Settlement Act 2004 are met. However, inappropriate aquaculture locations and unsustainable practices have the potential to compromise values and resources important to coastal whanau, hapu and iwi.

In 2008, a settlement of $97 million was made to Mâori for Crown obligations for aquaculture space that was approved between 1992 and 2004. NIWA operates a Mâori research and development unit, Te Kûwaha o Taihoro Nukurangi. The unit has a team of Mâori scientists who undertake research and provide consultancy services, based particularly aroundiwi with environmental and commercial issues.

### Prospects

In 2006, the New Zealand aquaculture industry published *The New Zealand Aquaculture Strategy*, setting itself an annual sales target of one billion NZ dollars by 2025. The strategy sets out ten areas of activity needed to achieve this target. The New Zealand Aquaculture Council has introduced a levy on aquaculture producers so this strategy can be implemented. In 2007, the New Zealand government responded to this industry initiative by releasing an aquaculture development strategy highlighting existing actions and proposing new initiatives. In addition, the government has offered additional funding around five key objectives, with the main focus on improving the implementation of the new 2004 regulations. At the end of 2008 there was a change in government, which announced that the aquaculture reforms will be overhauled, but reaffirmed the government commitment to the industry billion dollar target.

The New Zealand industry currently relies on low-value filter feeding shellfish (mussels and oysters) which are fast growing and relatively easy to culture. There is potential for the industry to diversify into higher value species such as pâua, kingfish and crayfish. These species need special food supplies and are more expensive to farm, but they command higher prices.

- The native blackfoot pâua, *Haliotis iris*, is a form of abalone. They are large sea snails which survive strong tidal surges by clinging to rocks using their large muscular foot. Wild pâua has been harvested since 1944, usually by skindivers,. Pâua aquaculture started in 1980, but has been slow moving beyond development. They are difficult to grow, and grow more slowly than salmon, mussels or oysters. Their larva and juveniles need to be grown separately. Most pâua farmers get juveniles from hatcheries, and feed them fresh kelp in land-based tanks. In Akaroa Harbour, one farmer grows pâua on plastic barrels tethered to buoys. In 2002, farmers produced five tonnes of pâua meat, worth $400,000.
- Another New Zealand pioneer pâua farmer cultivates blue pearls by placing some grit between the flesh of the pâua and its shell, where it acts as an irritant. The pâua responds by coating the grit with nacre (mother-of-pearl). This develops as a blue pearl.
- For some years there has been research on the best ways of growing the red seaweed, *Gigartina atropurpurea*, in New Zealand. Seaweed spores are grown on three metre strings at NIWA's Mahanga Bay aquaculture research facility, and are then transferred to a mussel farm in the Marlborough Sounds. If successful, a new seaweed growing industry could spread to the mussel farms in the Marlborough Sounds.

- Bluff oysters are harvested from the wild in Foveaux Strait. However, they breed more easily in Northland, and NIWA is examining their aquaculture possibilities.
- Shortfin and longfin eels have been trialled by NIWA. Established worldwide markets in cultured eels are worth over US$1 billion, and a decline in some stocks has opened up opportunities for New Zealand.
- The big-bellied seahorse is a native seahorse. Seahorses are valued aquarium fish. They are also used medicinally, particularly in traditional Chinese medicine. Wild seahorses have been over-harvested worldwide, opening markets to their aquaculture.
- A New Zealand sea sponge, *Mycale hentscheli*, which grows in Pelorus Sound, may hold the key to an anti-cancer drug. A substance produced by the sponges, might be used as a cancer-fighting drug. Victoria University and NIWA are working with Marlborough marine farmers to develop a method for growing the sponge on an existing mussel farm.

*Other prospects which are being researched or trialled include*

- European perch
- groper
- yellowtail kingfish, a native kingfish
- kina, a native sea urchin
- koura, a native freshwater lobster
- grass carp
- hâpuka, a native wreckfish
- sea cucumber
- rock lobster
- *Spongia manipulatus*, a native bath sponge
- giant kelp

## Timeline

- *Pre-European:* The indigenous Mâori undertake rudimentary aquaculture activities, such as placing suitable rocks into the intertidal settlement zones of oyster larvae. They are also thought to have transplanted abalone and other shellfish between different areas.
- *Early 20th century:* Salmon species are introduced to New Zealand as sport fish, but only the Chinook, or king salmon adapts to the environment.
- *1950s:* the Pacific oyster is introduced, possibly from a Japanese vessel hull or in their ballast water.

- *Early 1960s:* Dredge fisheries start operating in the north of the South Island and around the Hauraki Gulf. Within a few years they dredge these areas bare.
- *Late 1960s:* As a response to the collapse of the dredge fisheries, the aquaculture of New Zealand mussels begins.
- *1970s:* Farming of king salmon begins.
- *Late 1990s:* The aquaculture industry goes through a boom period, and demand for coastal space increases fivefold.
- *2002:* The government, in some disarray, imposes a moratorium on new marine farms while they attempt to develop better legislation aimed at dealing with the environmental demands of aquaculture and streamlining applications for marine and freshwater farms.
- *2005:* Parliament passes the Aquaculture Reform Act 2004, amending five existing acts so they can better cope with the environmental demands of aquaculture, and creating two new acts. However, the reform fails to streamline applications, and no further allocation of aquaculture space occurs over the next four years.
- *2006:* The New Zealand aquaculture industry publishes *The New Zealand Aquaculture Strategy*, setting itself an annual sales target of one billion NZ dollars by 2025.
- *2007:* The New Zealand government responds to the industry initiative by releasing an aquaculture development strategy highlighting existing actions and proposing new initiatives including funding incentives, mainly aimed at trying to action its reform legislation.
- *2008:* A settlement of $97 million is made to Mâori for Crown obligations for aquaculture space that was approved between 1992 and 2004.
- *2008:* The government changes and announces that the aquaculture reforms will be overhauled. It reaffirms the government commitment to the industry billion dollar target.

# 4

# Aquarium Management

An aquarium (plural *aquariums* or *aquaria*) is a vivarium consisting of at least one transparent side in which water-dwelling plants or animals are kept. Fishkeepers use aquaria to keep fish, invertebrates, amphibians, marine mammals, turtles, and aquatic plants. The term combines the Latin root *aqua*, meaning water, with the suffix *-arium*, meaning "a place for relating to".

An aquarist owns fish or maintains an aquarium, typically constructed of glass or high strength acrylic plastic. Cuboid aquaria are also known as fish tanks or simply tanks, while bowl-shaped aquaria are also known as fish bowls. Size can range from a small glass bowl to immense public aquaria. Specialised equipment maintains appropriate water quality and other characteristics suitable for the aquarium's residents.

## History and Popularisation

### Antiquity

In the Roman Empire, the first fish to be brought indoors was the sea barbel, which was kept under guest beds in small tanks made of marble. Introduction of glass panes around the year 50 allowed Romans to replace one wall of marble tanks, improving their view of the fish.

In 1369, the Chinese Emperor, HóngwÔ, established a porcelain company that produced large porcelain tubs for maintaining goldfish; over time, people produced tubs that approached the shape of modern fish bowls. Leonhard Baldner, who wrote *Vogel-, Fisch- und Tierbuch*

(Bird, Fish, and Animal Book) in 1666, maintained weather loaches and newts.

### Nineteenth Century

In 1836, soon after his invention of the Wardian case, Dr. Nathaniel Bagshaw Ward proposed to use his tanks for tropical animals. In 1841 he did so, though only with aquatic plants and toy fish. However, he soon housed real animals. In 1838, Félix Dujardin noted owning a saltwater aquarium, though he did not use the term. In 1846, Anne Thynne maintained stony corals and seaweed for almost three years, and was cred as the creator of the first balanced marine aquarium in London. At about the same time, Robert Warington experimented with a 13-gallon container, which contained goldfish, eelgrass, and snails, creating one of the first stable aquaria. He published his findings in 1850 in the Chemical Society's journal.

The keeping of fish in an aquarium became a popular hobby and spread quickly. In the United Kingdom, it became popular after ornate aquaria in cast iron frames were featured at the Great Exhibition of 1851. In 1853, the first large public aquarium opened in the London Zoo and came to be known as the Fish House. Philip Henry Gosse was the first person to actually use the word "aquarium", opting for this term (instead of "aquatic vivarium" or "aqua-vivarium") in 1854 in his book *The Aquarium: An Unveiling of the Wonders of the Deep Sea*. In this book, Gosse primarily discussed saltwater aquaria. In the 1850s, the aquarium became a fad in the United Kingdom. Tank designs and techniques for maintaining water quality were developed by Warington, later cooperating with Gosse until his critical review of the tank water composition. Edward Edwards developed these glass-fronted aquaria in his 1858 patent for a "dark-water-chamber slope-back tank", with water slowly circulating to a reservoir beneath.

Germans soon rivalled the British in their interest. In 1854, an anonymous author had two articles published about the saltwater aquaria of the United Kingdom: *Die Gartenlaube* (The Garden House) entitled *Der Ocean auf dem Tische*. However, in 1856,*Der See im Glase* (The Lake in a Glass) was published, discussing freshwater aquaria, which were much easier to maintain in landlocked areas. During the 1870s, some of the first aquarist societies were appearing in Germany. The United States soon followed. Published in 1858, Henry D. Butler's *The Family Aquarium* was one of the first books written in the United States solely about the aquarium. According to the July issue of *The North American Review* of the same year, William Stimson may have owned some of the first functional aquaria, and had as many as seven or eight. The first aquarist society in the United States was founded in New York City in 1893, followed by others. The *New York Aquarium Journal*, first published in October 1876, is considered to be the world's first aquarium magazine.

In the Victorian era in the United Kingdom, a common design for the home aquarium was a glass front with the other sides made of wood (made watertight with a pitch coating). The bottom would be made of slate and heated from below. More advanced systems soon began to be introduced, along with tanks of glass in metal frames. During the latter half of the 19th century, a variety of aquarium designs were explored, such as hanging the aquarium on a wall, mounting it as part of a window, or even combining it with a birdcage.

### *Twentieth Century*

Circa 1908, the first mechanical aquarium air pump was invented, powered by running water, instead of electricity. The introduction of the air pump into the hobby is considered by several historians of the hobby to be a pivotal moment in its development.

Aquaria became more widely popular as houses had an electricity supply after World War I. Electricity allowed artificial lighting as well as aeration, filtration, and heating of the water. Initially, amateur aquarists kept native fish (with the exception of goldfish); the availability of exotic species from overseas further increased the popularity of the aquarium. Jugs made from a variety of materials were used to import fish from overseas, with a bicycle foot pump for aeration. Plastic shipping bags were introduced in the 1950s, making it easier to ship fish. The eventual availability of air freight allowed fish to be successfully imported from distant regions. In the 1960s metal frames made marine aquaria almost impossible due to corrosion, but the development of tar and silicone sealant allowed the first all-glass aquaria made by Martin Horowitz in Los Angeles, CA. The frames remained, however, though purely for aesthetic reasons.

In the United States, as of 1996, aquarium keeping is the second-most popular hobby after stamp collecting. In 1999 it was estimated that over nine million U.S. households own an aquarium. Figures from the 2005/2006 APPMA National Pet Owners Survey report that

Americans own approximately 139 million freshwater fish and 9.6 million saltwater fish. Estimates of the numbers of fish kept in aquaria in Germany suggest at least 36 million. The hobby has the strongest following in Europe, Asia, and North America. In the United States, 40 percent of aquarists maintain two or more tanks.

## Design

### *Materials*

Most aquaria consist of glass panes bonded together by silicone, with plastic frames that are attached to the upper and lower edges for decoration. The glass aquarium is standard for sizes up to about 1,000 litres (260 US gal). However, glass as a material is brittle and has very little give before fracturing, though generally the sealant fails first. Aquaria come in a variety of shapes such as cuboid, hexagonal, angled to fit in a corner (L-shaped), and bow-front (the front side curves outwards). Fish bowls are generally either made of plastic or glass, and are either spherical or some other round configuration in shape. The very first modern aquarium made of glass was developed in the 19th century by Robert Warrington. During the Victorian age it was common for glass aquaria to have slate or steel bottoms, which allowed the aquaria to be heated underneath with an open flame heat source. The aquaria in those days had the glass panels attached with metal frames and sealed with putty. These metal framed aquaria were still available on the market until the mid 1960s when the modern, silicone-sealed style replaced them. Acrylic tanks were not generally available to the public until the 1970s.

Although glass aquaria are usually preferred by aquarists over the acrylic ones because of their resistance to scratching and much more accessible price, they come with several disadvantages. Not only are they not as crack resistant as acrylic tanks but they are also nearly two times heavier than the latter. They also provide less insulation than acrylic aquaria and do not come in as many interesting shapes. Many aquarists or beginners who want to get fish as pets find it particularly onerous that many online suppliers will not ship glass aquaria because of the high potential for cracking and the high weight, which increases the cost of shipping. However, glass tanks are more convenient for other aquarists because unlike acrylic, glass does not yellow over time, and also because glass tanks do not need as much support as acrylic aquaria.

Even though the price is one of the main considerations for aquarists when deciding which of these two types of aquaria to purchase, when it comes to very large tanks the price difference tends to disappear.

Acrylic aquaria are also available and are the primary competitor with glass. Acrylic aquaria are stronger than glass, and much lighter. Acrylic-soluble cements are used to directly fuse acrylic together (as opposed to simply sealing the seam). Acrylic allows for the formation of unusual shapes, such as the hexagonal tank. Compared to glass, acrylics are easy to scratch; but unlike glass, it is possible to polish out scratches in acrylic.

Laminated glass is sometimes used, which combines the advantages of both glass and acrylic. Large aquaria might instead use stronger materials such as fiberglass-reinforced plastics. However, this material is not transparent. Reinforced concrete is used for aquaria where weight and space are not factors. Concrete must be coated with a waterproof layer to prevent the water from breaking down the concrete as well as preventing contamination of the water by the concrete.

### *Styles*

Aquariums have been fashioned into coffee tables, sinks, and even toilets. Another such example is the MacQuarium, an aquarium made from the shell of an Apple Macintosh computer. In recent years, elaborate custom-designed home aquariums costing hundreds of thousands of dollars have become status symbols—according to *The New York Times*, "among people of means, a dazzling aquarium is one of the last sure fire ways to impress their peers."

A kreisel tank is a circular aquarium designed to hold delicate animals such as jellyfish. These aquariums provide slow, circular water flow with a bare minimum of interior hardware, to prevent delicate animals from becoming injured by pumps or the tank itself. Originally a German design (*kreisel* means spinning top), the tank has no sharp corners, and keeps the housed animals away from the plumbing. Water moving into the tank gives a gentle flow that keeps the inhabitants suspended, and water leaving the tank is covered by a delicate screen that prevents the inhabitants from getting stuck. There are several types of kreisel tanks. In a true kreisel, a circular tank has a circular, submerged lid. Pseudokreisels have a curved bottom surface and a flat top surface, similar to the shape of either a "U" or a semicircle. Stretch kreisels or Langmuir kreisels are a "double gyre" kreisel design, where the tank length is at least twice the height. Using two downwelling inlets on both sides of the tank lets gravity create two gyres in the tank.

A single downwelling inlet may be used in the middle as well. The top of a stretch kreisel may be open or closed with a lid. There may also be screens about midway down the sides of the tank, or at the top on the sides. It is possible to combine these designs; a circular shaped tank is used without a lid or cover, and the surface of the water acts as the continuation of circular flow. It is now possible to start a jellyfish aquarium at home as easily as a regular fish tank.

Another popular setup is the biotope aquarium. A biotope aquarium is a recreation of a specific natural environment. Some of the most popular biotopes (to name only a few) are the Amazon river, Rio Negro River, Lake Malawi, Lake Tanganyika, and Lake Victoria. The fish, plants, substrate, rocks, wood, and any other component of the display should match that of the natural environment. It can be a real challenge to recreate such environments and most "true" biotopes will only have a few species of fish (if not only one) and invertebrates.

### *Aquarium Size and Volume*

An aquarium can range from a small glass bowl containing less than 1 litre (2.1 US pt) of water to immense public aquaria that house entire ecosystems such as kelp forests. Relatively large home aquaria resist rapid fluctuations of temperature and pH, allowing for greater system stability.

Unfiltered bowl-shaped aquaria are now widely regarded as unsuitable for most fish. Advanced alternatives are now available. In order to keep water conditions at suitable levels, aquariums should contain at least two forms of filtration: biological and mechanical. Chemical filtration should also be considered under some circumstances for optimum water quality. Chemical filtration is frequently achieved via activated carbon, to filter medications, tannins, and/or other known impurities from the water.

Reef aquaria under 100 litres (26 US gal) have a special place in the aquarium hobby; these aquaria, termed nano reefs (when used in reefkeeping), have a small water volume.

Practical limitations, most notably the weight of water (1 kilogram per litre (8.345 lb/U.S. gal)) and internal water pressure (requiring thick glass siding) of a large aquarium, restrict most home aquaria to a maximum of around 1 cubic metre in volume (1000 L, weighing 1,000 kg or 2,200 lb). Some aquarists, however, have constructed aquaria of many thousands of litres.

Public aquariums designed for exhibition of large species or environments can be dramatically larger than any home aquarium. The Georgia Aquarium, for example, features an individual aquarium of 6,300,000 US gallons (24,000 $m^2$).

### Components

The typical hobbyist aquarium includes a filtration system, an artificial lighting system, and a heater or chiller depending on the aquarium's inhabitants. Many aquaria incorporate a hood, to decrease evapouration and prevent fish from leaving the aquarium (and anything else from entering the aquarium). They also often hold lights.

Combined biological and mechanical aquarium filtration systems are common. These either convert ammonia to nitrate (removing nitrogen at the expense of aquatic plants), or to sometimes remove phosphate. Filter media can house microbes that mediate nitrification. Filtration systems are the most complex component of home aquaria.

Aquarium heaters combine a heating element with a thermostat, allowing the aquarist to regulate water temperature at a level above that of the surrounding air, whereas coolers and chillers (refrigeration devices) are for use anywhere, such as cold water aquaria, where the ambient room temperature is above the desired tank temperature. Thermometres used include glass alcohol thermometres, adhesive external plastic strip thermometres, and battery-powered LCD thermometres. In addition, some aquarists use air pumps attached to airstones or water pumps to increase water circulation and supply adequate gas exchange at the water surface. Wave-making devices have also been constructed to provide wave action.

An aquarium's physical characteristics form another aspect of aquarium design. Size, lighting conditions, density of floating and rooted plants, placement of bog-wood, creation of caves or overhangs, type of substrate, and other factors (including an aquarium's positioning within a room) can all affect the behaviour and survival of tank inhabitants. An aquarium can be placed on an aquarium stand. Because of the weight of the aquarium, a stand must be strong as well as level. A tank that is not level may distort, leak, or crack. These are often built with cabinets to allow storage, available in many styles to match room decor. Simple metal tank stands are also available. Most aquaria should be placed on polystyrene to cushion any irregularities on the underlying surface or the bottom of the tank itself. However, some tanks have an underframe making this unnecessary.

### Aquarium Maintenance

Large volumes of water enable more stability in a tank by diluting effects from death or contamination events that push an aquarium away from equilibrium. The bigger the tank, the easier such a systemic shock is to absorb, because the effects of that event are diluted.

For example, the death of the only fish in a three U.S. gallon tank (11 L) causes dramatic changes in the system, while the death of that same fish in a 100 U.S. gallon (400 L) tank with many other fish in it represents only a minor change. For this reason, hobbyists often favour larger tanks, as they require less attention.

Several nutrient cycles are important in the aquarium. Dissolved oxygen enters the system at the surface water-air interface. Similarly, carbon dioxide escapes the system into the air. The phosphate cycle is an important, although often overlooked, nutrient cycle. Sulphur, iron, and micronutrients also cycle through the system, entering as food and exiting as waste. Appropriate handling of the nitrogen cycle, along with supplying an adequately balanced food supply and considered biological loading, is enough to keep these other nutrient cycles in approximate equilibrium.

An aquarium must be maintained regularly to ensure that the fish are kept healthy. Daily maintenance consists of checking the fish for signs of stress and disease. Also, aquarists must make sure that the water has a good quality and it is not cloudy or foamy and the temperature of the water is appropriate for the particular species of fish that live in the aquarium.

Typical weekly maintenance includes changing around 10-20% of the water while cleaning the gravel, or other substrate if the aquarium has one. A good habit is to remove the water being replaced by "vacuuming" the gravel with suitable implements, as this will eliminate uneaten foods and other residues that settle on the substrate. Tap water is not considered to be safe for fish to live in because it contains chemicals that harm the fish, so any tap water used must be treated with a suitable water conditioner, such as a product which removes chlorine and chloramine, and neutralises any heavy metals present. The water parameters must be checked both in the tank and in the replacing water, to make sure they are suitable for the species of fish kept.

### Water Conditions

The solute content of water is perhaps the most important aspect of water conditions, as total dissolved solids and other constituents dramatically impact basic water chemistry, and therefore how organisms interact with their environment. Salt content, or salinity, is the most basic measure of water conditions. An aquarium may have freshwater (salinity below 500 parts per million), simulating a lake or river environment; brackish water (a salt level of 500 to 30,000 PPM), simulating environments lying between fresh and salt, such as estuaries; and salt water or seawater (a salt level of 30,000 to 40,000 PPM), simulating an ocean environment. Rarely, higher salt concentrations are maintained in specialised tanks for raising brine organisms.

Saltwater is typically alkaline, while the pH (alkalinity or acidicity) of fresh water varies more. Hardness measures overall dissolved mineral content; hard or soft water may be preferred. Hard water is usually alkaline, while soft water is usually neutral to acidic. Dissolved organic content and dissolved gases content are also important factors.

Home aquarists typically use tap water supplied through their local water supply network to fill their tanks. Straight tap water cannot be used in localities that pipe chlorinated water. In the past, it was possible to "condition" the water by simply letting the water stand for a day or two, which allows the chlorine time to dissipate. However, chloramineis now used more often and does not leave the water as readily. Additives formulated to remove chlorine or chloramine are often all that is needed to make the water ready for aquarium use. Brackish or saltwater aquaria require the addition of a commercially available mixture of salts and other minerals.

Some aquarists modify water's alkalinity, hardness, or dissolved content of organics and gases, before adding it to their aquaria. This can be accomplished by additives, such as sodium bicarbonate, to raise pH. Some aquarists filter or purify their water through deionisation or reverse osmosis prior to using it. In contrast, public aquaria with large water needs often locate themselves near a natural water source (such as a river, lake, or ocean) to reduce the level of treatment. Some hobbyists use an algae scrubber to filter the water naturally.

Water temperature determines the two most basic aquarium classifications: tropical versus cold water. Most fish and plant species tolerate only a limited temperature range; tropical aquaria, with an average temperature of about 25 °C (77 °F), are much more common. Cold water aquaria are for fish that are better suited to a cooler environment. More important than the range is consistency; most organisms are not accustomed to sudden changes in temperatures, which can cause shock and lead to disease. Water temperature can be regulated with a thermostat and heater (or cooler).

Water movement can also be important in simulating a natural ecosystem. Aquarists may prefer anything from still water up to swift currents, depending on the aquarium's inhabitants. Water movement can be controlled via aeration from air pumps, powerheads, and careful design of internal water flow (such as location of filtration system points of inflow and outflow).

### Nitrogen Cycle

Of primary concern to the aquarist is management of the waste produced by an aquarium's inhabitants. Fish, invertebrates, fungi, and some bacteria excrete nitrogen waste in the form of ammonia (which converts to ammonium, in acidic water) and must then pass through the nitrogen cycle. Ammonia is also produced through the decomposition of plant and animal matter, including fecal matter and other detritus. Nitrogen waste products become toxic to fish and other aquarium inhabitants at high concentrations.

### The Process

A well-balanced tank contains organisms that are able to metabolise the waste products of other aquarium residents. This process is known in the aquarium hobby as the nitrogen cycle. Bacteria known as nitrifiers (genus *Nitrosomonas*) metabolise nitrogen waste. Nitrifying bacteria capture ammonia from the water and metabolise it to producenitrite. Nitrite is toxic to fish in high concentrations.

Another type of bacteria (genus *Nitrospira*) converts nitrite into nitrate, a less toxic substance. (*Nitrobacter*bacteria were previously believed to fill this role. While biologically they could theoretically fill the same niche as *Nitrospira*, it has recently been found that *Nitrobacter* are not present in detectable levels in established aquaria, while *Nitrospira* are plentiful.) However, commercial products sold as kits to "jump start" the nitrogen cycle often still contain *Nitrobacter*.

In addition to bacteria, aquatic plants also eliminate nitrogen waste by metabolising ammonia and nitrate. When plants metabolise nitrogen compounds, they remove nitrogen from the water by using it to build biomass that decays more slowly than ammonia-driven plankton already dissolved in the water.

### *Maintaining the Nitrogen Cycle*

What hobbyists call the nitrogen cycle is only a portion of the complete cycle: nitrogen must be added to the system (usually through food provided to the tank inhabitants), and nitrates accumulate in the water at the end of the process, or become bound in the biomass of plants. The aquarium keeper must remove water once nitrate concentrations grow, or remove plants which have grown from the nitrates.

Hobbyist aquaria often do not have sufficient bacteria populations to adequately denitrify waste. This problem is most often addressed through two filtration solutions: Activated carbon filters absorb nitrogen compounds and other toxins, while biological filters provide a medium designed to enhance bacterial colonisation. Activated carbon and other substances, such as ammonia absorbing resins, stop working when their pores fill, so these components have to be replaced regularly.

New aquaria often have problems associated with the nitrogen cycle due to insufficient beneficial bacteria. Therefore fresh water has to be matured before stocking them with fish. There are three basic approaches to this: the "fishless cycle", the "silent cycle" and "slow growth".

In a fishless cycle, small amounts of ammonia are added to an unpopulated tank to feed the bacteria. During this process, ammonia, nitrite, and nitrate levels are tested to monitor progress. The "silent" cycle is basically nothing more than densely stocking the aquarium with fast-growing aquatic plants and relying on them to consume the nitrogen, allowing the necessary bacterial populations time to develop. According to anecdotal reports, the plants can consume nitrogenous waste so efficiently that ammonia and nitrite level spikes seen in more traditional cycling methods are greatly reduced or disappear. "Slow growth" entails slowly increasing the population of fish over a period of 6 to 8 weeks, giving bacteria colonies time to grow and stabilise with the increase in fish waste. This method is usually done with a small starter population of hardier fish which can survive the ammonia and nitrite spikes, whether they are intended to be permanent residents or to be traded out later for the desired occupants.

The largest bacterial populations are found in the filter, where is high water flow and plentiful surface available for their growth, so effective and efficient filtration is vital. Sometimes, a vigorous cleaning of the filter is enough to seriously disturb the biological balance of an aquarium. Therefore, it is recommended to rinse mechanical filters in an outside bucket of aquarium water to dislodge organic materials that contribute to nitrate problems, while preserving bacteria populations. Another safe practice consists of cleaning only half of the filter media during each service, or using two filters, only one of which is cleaned at a time.

### *Biological Loading*

Biological load is a measure of the burden placed on the aquarium ecosystem by its inhabitants. High biological loading presents a more complicated tank ecology, which in turn means that equilibrium is easier to upset. Several fundamental constraints on biological loading depend on aquarium size. The water's surface area limits oxygen intake. The bacteria population depends on the physical space they have available to colonise. Physically, only a limited size and number of plants and animals can fit into an aquarium while still providing room for movement. Biologically, biological loading refers to the rate of biological decay in proportion to tank volume. Adding plants to an aquarium will sometimes help greatly with taking up fish waste as plant nutrients. Although an aquarium can be overloaded with fish, an excess of plants is unlikely to cause harm.

### *Calculating Capacity*

Limiting factors include the oxygen availability and filtration processing. Aquarists have rules of humb to estimate the number of fish that can be kept in an aquarium. The examples below are for small freshwater fish; larger freshwater fishes and most marine fishes need much more generous allowances.

- 3 cm of *adult* fish length per 4 litres of water (i.e., a 6 cm-long fish would need about 8 litres of water).
- 1 cm of *adult* fish length per 30 square centimetres of surface area.
- 1 inch of *adult* fish length per US gallon of water.
- 1 inch of *adult* fish length per 12 square inches of surface area.

Experienced aquarists warn against applying these rules too strictly because they do not consider other important issues such as growth rate, activity level, social behaviour, filtration capacity, total biomass of plant life, and so on. Establishing maximum capacity is often a matter of slowly adding fish and monitoring water quality over time, following a trial and error approach.

### *Other Factors Affecting Capacity*

One variable is differences between fish. Smaller fish consume more oxygen per gram of body weight than larger fish. Labyrinth fish can breathe atmospheric oxygen and do not need as much surface area (however, some of these fish are territorial, and do not appreciate crowding). Barbs also require more surface area than tetras of comparable size.

Oxygen exchange at the surface is an important constraint, and thus the surface area of the aquarium matters. Some aquarists claim that a deeper aquarium holds no more fish than a shallower aquarium with the same surface area. The capacity can be improved by surface movement and water circulation such as through aeration, which not only improves oxygen exchange, but also waste decomposition rates.

Waste density is another variable. Decomposition in solution consumes oxygen. Oxygen dissolves less readily in warmer water; this is a double-edged sword since warmer temperatures make fish more active, so they consume more oxygen.

In addition to bioload/chemical considerations, aquarists also consider the mutual compatibility of the fish. For instance, predatory fish are usually not kept with small, passive species, and territorial fish are often unsuitable tankmates for shoaling species. Furthermore, fish tend to fare better if given tanks conducive to their size. That is, large fish need large tanks and small fish can do well in smaller tanks. Lastly, the tank can become overcrowded without being overstocked. In other words, the aquarium can be suitable with regard to filtration capacity, oxygen load, and water, yet still be so crowded that the inhabitants are uncomfortable.

### Aquarium Classifications

From the outdoor ponds and glass jars of antiquity, modern aquaria have evolved into a wide range of specialised systems. Individual aquaria can vary in size from a small bowl large enough for only a single small fish, to the huge public aquaria that can simulate entire marine ecosystems.

One way to classify aquaria is by salinity. Freshwater aquaria are the most popular due to their lower cost. More expensive and complex equipment is required to set up and maintain a marine aquaria. Marine aquaria frequently feature a diverse range of invertebrates in addition to species of fish. Brackish water aquaria combine elements of both marine and freshwater fish keeping. Fish kept in brackish water aquaria generally come from habitats with varying salinity, such as mangrove swamps and estuaries. Subtypes exist within these types, such as the reef aquarium, a typically smaller marine aquarium that houses coral.

Another classification is by temperature range. Many aquarists choose a tropical aquarium because tropical fish tend to be more colourful. However, the coldwater aquariumis also popular, which is mainly restricted to goldfish, but can include fish from temperate areas worldwide and native fish keeping.

Aquaria may be grouped by their species selection. The community tank is the most common today, where several non-aggressive species live peacefully. In these aquaria, the fish, invertebrates, and plants probably do not originate from the same geographic region, but tolerate similar water conditions. Aggressive tanks, in contrast, house a limited number of species that can be aggressive toward other fish, or are able to withstand aggression well. Most marine tanks and tanks housing cichlids have to take the aggressiveness of the desired species into account when stocking. Specimen tanks usually only house one fish species, along with plants, perhaps ones found in the fishes' natural environment and decorations simulating a natural ecosystem. This type is useful for fish that cannot coexist with other fish, such as the electric eel, as an extreme example. Some tanks of this sort are used simply to house adults for breeding.

Ecotype, ecotope, or biotope aquaria is another type based on species selection. In it, an aquarist attempts to simulate a specific natural ecosystem, assembling fish, invertebrate species, plants, decorations and water conditions all found in that ecosystem. These biotope aquaria are the most sophisticated hobby aquaria; public aquaria use this approach whenever possible. This approach best simulates the experience of observing in the wild. It typically serves as the healthiest possible artificial environment for the tank's occupants.

### Public Aquaria

Most public aquarium facilities feature a number of smaller aquaria, as well those too large for home aquarists. The largest tanks hold millions of gallons of water and can house large species, including sharks or beluga whales.

Dolphinaria are specifically for dolphins. Aquatic and semiaquatic animals, including otters and penguins, may also be kept by public aquaria. Public aquaria may also be included in larger establishments such as a marine mammal park or a marine park.

### Virtual Aquariums

A virtual aquarium is a computer program which uses 3D graphics to reproduce an aquarium on a personal computer. The swimming fish are rendered in real time, while the background of the tank is usually static. Objects on the floor of the tank may be mapped in simple planes so that the fish may appear to swim both in front and behind them, but a relatively simple 3D map of the general shape of such objects may be used to allow the light and ripples on the surface of the water to cast realistic shadows.

Bubbles and water noises are common for virtual aquariums, which are often used as screensavers.

The number of each type of fish can usually be selected, often including other animals like starfish, jellyfish, seahorses, and even sea turtles. Other objects found in an aquarium can also be added and rearranged on some software, like treasure chests and giant clams that open and close with air bubbles, or a bobbing diver. There are also usually features that allow the user to tap on the glass or put food in the top, both of which the fish will react to. Some also have the ability to allow the user to  fish and other objects to create new varieties.

## Aquascaping

Aquascaping is the craft of arranging aquatic plants, as well as rocks, stones, cavework, or driftwood, in an aesthetically pleasing manner within an aquarium—in effect, gardening under water. Aquascape designs include a number of distinct styles, including the garden-like Dutch style and the Japanese-inspired nature style.

Typically, an aquascape houses fish as well as plants, although it is possible to create an aquascape with plants only, or with rockwork or other hardscape and no plants.

Although the primary aim of aquascaping is to create an artful underwater landscape, the technical aspects of aquatic plant maintenance must also be taken into consideration.

Many factors must be balanced in the closed system of an aquarium tank to ensure the success of an aquascape.

These factors include filtration, maintaining carbon dioxide at levels sufficient to support photosynthesis underwater, substrate and fertilization, lighting, and algae control.

Aquascape hobbyists trade plants, conduct contests, and share photographs and information via the internet. The United States-based Aquatic Gardeners Association has about 1,200 members.

## Designs

### *Dutch Style*

The Dutch aquarium employs a lush arrangement in which multiple types of plants having diverse leaf colours, sizes, and textures are displayed much as terrestrial plants are shown in a flower garden.

This style was developed in the Netherlands starting in the 1930s, as freshwater aquarium equipment became commercially available. It emphasizes plants located on terraces of different heights, and frequently omits rocks and driftwood.

Linear rows of plants running left-to-right are referred to as "Dutch streets". Although many plant types are used, one typically sees neatly trimmed groupings of plants with fine, feathery foliage, such as *Limnophila aquatica* and various types of *Hygrophila*, along with the use of red-leaved *Alternanthera reineckii*, *Ammania gracilis*, and assorted *Rotala* for colour highlights. More than 80% of the aquarium floor is covered with plants, and little or no substrate is left visible. Tall growing plants that cover the back glass originally served the purpose of hiding bulky equipment behind the tank.

### Nature Style

A contrasting approach is the "nature aquarium" or Japanese style, introduced in the 1990s by Takashi Amano. Amano's three-volume series, *Nature Aquarium World,* sparked a wave of interest in aquarium gardening, and he has been cited as having "set a new standard in aquarium management".

Amano's compositions draw on Japanese gardening techniques that attempt to mimic natural landscapes by the asymmetrical arrangement of masses of relatively few species of plants, and carefully selected stones or driftwood. The objective is to evoke a landscape in miniature, rather than a colourful garden.

This style draws particularly from the Japanese aesthetic concepts of *Wabi-sabi,* which focuses on transience and minimalism as sources of beauty, and *Iwagumi,* which sets rules governing rock placement.

In the *Iwagumi* system, the *Oyaishi* or main stone, is placed slightly off-centre in the tank, and *Soeishi* or accompanying stones, are grouped near it, while *Fukuseki*, or secondary stones, are arranged in subordinate positions.

The location of the focal point of the display, determined largely by the asymmetric placement of the *Oyaishi*, is considered important, and follows ratios that reflect Pythagorean tuning. Plants with small leaves, such as *Eleocharis acicularis*, *Glossostigma elatinoides*, *Hemianthus callitrichoides*, *Riccia fluitans*, small aquatic ferns, and Java moss (*Versicularia dubyana*or *Taxiphyllum barbieri*) are usually emphasized, with more limited colours than in the Dutch style, and the hardscape is not completely covered. Fish, or freshwater shrimp such as *Caridina multidentata* and *Neocaridina heteropoda*, are usually selected to complement the plants and control algae.

### *Jungle Style*

Some hobbyists also refer to a "jungle" (or "wild jungle") style, separate from either the Dutch or nature styles, and incorporating some of the features of them both. Bold, coarser leaf shapes, such as *Echinodorus bleheri*, are used to provide a wild, untamed appearance.

### *Biotopes*

The styles above often combine plant and animal species based on the desired visual impact, without regard to geographic origin. Biotope aquascapes are designed instead to replicate exactly a particular aquatic habitat at a particular geographic location, and not necessarily to provide a garden-like display. Plants and fish need not be present at all, but if they are, they must match what would be found in nature in the habitat being represented, as must any gravel and hardscape, and even the chemical composition of the water.

### *Paludariums*

In a paludarium, part of the aquarium is underwater, and part is above water. Substrate is built up so that some "land" regions are raised above the waterline, and the tank is filled with water only part way.

This allows plants, such as *Cyperus alternifolius* and *Spathiphyllum wallisii*, as well as various *Anubias* and some bromeliads, to grow emersed, with their roots underwater but their tops in the air, as well as completely submersed. In some configurations, plants that float on the surface of the water, such as *Eichhornia crassipes* and *Pistia stratiotes*, can be displayed to full advantage. Unlike other aquarium setups, paludariums are particularly well-suited to keeping amphibians.

### *Saltwater Reefs*

Dutch and nature style aquascapes are traditionally freshwater systems. In contrast, relatively few ornamental plants can be grown in a saltwater aquarium. Saltwater aquascaping typically centres, instead, on mimicking a reef.

An arrangement of live rock forms the main structure of this aquascape, and it is populated by corals and other marine invertebrates as well as coralline algae, which together serve much the same aesthetic role as freshwater plants.

Lighting plays a particularly significant role in the reef aquascape. Many corals, as well as tridacnid clams, contain symbiotic fluorescent algae-likeprotozoa called zooxanthellae. By providing intense lighting supplemented in the ultraviolet wavelengths, reef aquarists not only support the health of these invertebrates, but also elicit particularly bright colours emitted by the fluorescent microorganisms.

### Techniques

In addition to design, freshwater aquascaping also requires specific methods to maintain healthy plants underwater. Plants are often trimmed to obtain the desired shape, and they can be positioned by tying them in place inconspicuously with thread. Most serious aquascapers use aquarium-safe fertilizers, commonly in liquid or tablet form, to help the plants fill out more rapidly. Some aquarium substrates containing laterite also provide nutrients.

It is also necessary to support photosynthesis, by providing light and carbon dioxide. A variety of lighting systems may be used to produce the full spectrum of light, usually at 2–4watts per gallon (0.5–1 watts per litre). Lights are usually controlled by a timer that allows the plants to be acclimated to a set cycle. Depending on the number of plants and fish, the aquascape may also require carbon dioxide supplementation.

This can be accomplished with a simple homemade system, using a soda bottle filled with yeast, warm water, and sugar, and connected to an airstone in the aquarium, or with a pressurized $CO_2$ tank that injects a set amount of carbon dioxide into the aquarium water.

Algae (including cyanobacteria, as well as true algae) is considered distracting and unwanted in aquascaping, and is controlled in several ways. One is the use of animals that consume algae, such as some fish (notably cyprinids of the genera *Crossocheilus* and *Gyrinocheilus*, and catfish of the genera *Ancistrus*, *Hypostomus*, and *Otocinclus*), shrimp, orsnails, to clean the algae that collects on the leaves. A second is using adequate light and $CO_2$ to promote rapid growth of desired plants, while controlling nutrient levels, to ensure that the plants utilise all fertilizer without leaving nutrients to support algae.

Although serious aquascapers often use a considerable amount of equipment to provide lighting, filtration, and $CO_2$ supplementation to the tank, some hobbyists choose instead to maintain plants with a minimum of technology, and some have reported success in producing lush plant growth this way. This approach, sometimes called the "natural planted tank" and popularised by Diana Walstad, can include the use of soil in place of aquarium gravel, the elimination of $CO_2$ apparatus and most filtration, and limited lighting. Instead, only a few fish are kept, to limit the quantity of fish waste, and the plants themselves are used to perform the water-cleansing role typically played by aquarium filters, by utilising what fish waste there is as fertilizer.

### Contests

Early Dutch hobbyists began the practice of aquascape contests, with over 100 local clubs. Judges had to go through about three years of training and pass examinations in multiple disciplines in order to qualify. This competition continues to be held every year, under the auspices of the National Aquarium Society.

There are three rounds, beginning with contests in local clubs. First-place local winners are entered in the second round, held in fifteen *districtkeuring* (districts). The winners at that level are then entered in the third round, which is the national championship.

In the Dutch contest, the focus is not only on composition, but also on the biological well-being of the aquarium's inhabitants. Most points are, in fact, awarded for such biological criteria as fish health, plant health, and water quality.

Unlike contests in other countries, the judges travel to each contestant's home to evaluate the tank, where they measure the water parameters themselves.

The Aquatic Gardeners Association, based in the United States, Aqua Design Amano, based in Japan, and Aquatic Scapers Europe, based in Germany, also conduct annual freshwater aquascaping contests. Entries from around the world are submitted as photographs and explanatory text online.

The Aquatic Gardeners Association contest is judged based on:

- overall impression (35 points),
- composition, balance, use of space and use of colour (30 points),
- selection and use of materials (20 points), and
- viability of aquascape (15 points).

There are also smaller contests conducted by Acuavida in Spain, and by the Greek Aquarist's Club.

### *Public Aquariums*

Large public aquariums sometimes use aquascaping as part of their displays. As early as the 1920s, the New York Aquarium included a moray eel display tank that was decorated with calcareous tufa rock, arranged to resemble a coral reef, and supporting some stony corals and sea fans. Because they typically present wildlife from a particular habitat, modern day displays are often created to be biologically accurate biotopes.

## Association of Zoos and Aquariums

The Association of Zoos and Aquariums (previously American Zoo and Aquarium Association, and originally American Association of Zoological Parks and Aquariums) was founded in 1924 and is a nonprofit organisation dedicated to the advancement of zoos and public aquariums in the areas of conservation, education, science, and recreation. The AZA headquarters is located in Silver Spring, Maryland.

### *History*

In October 1924 the American Association of Zoological Parks and Aquariums (AAZPA) was formed as an affiliate of the American Institute of Park Executives (AIPE). In 1966, the AAZPA became a professional branch affiliate of the newly formed National Recreation and Park Association (NRPA which absorbed the AIPE).

In the fall of 1971, the AAZPA membership voted to become an independent association and, in January 1972, it was chartered as the AAZPA with its executive office located in Wheeling, West Virginia within the Oglebay Park Good Zoo. In January 1994, the shorter name American Zoo and Aquarium Association (AZA) was adopted.

In 1998, there were 134 million visits to North American zoos and aquariums. Ten years later, in 2008, there were 175 million visits to AZA-accred zoos and aquariums.

### *Activities*

The organisation is active in institution accration, animal care initiatives, education and conservation programs, collaborative research and political lobbying in order to achieve this goal. It serves as an accring body for zoos and aquariums and ensures that accred facilities meet higher standards of animal care than are required by law. Institutions are evaluated every five years in order to ensure standards are met and to maintain accration. The association also facilitates both Species Survival Plans and Population Management Plans, which serve to sustainably manage genetically diverse captive populations of various animal species. In addition, the association sponsors the scientific journal *Zoo Biology*.

### Annual Report on Conservation and Science

The association has established a computerised database called the *Annual Report on Conservation and Science*. It provides a model for a broader database to help track research projects worldwide. The database can be searched by key word, name of researcher, topic, country or region, name of institution, conservation program title, name of cooperating institution (including governmental agencies and non-governmental organisations, colleges or universities, and non-member zoos and aquariums), type of research, or date.

In 2000-2001, member institutions reported that they participated in over 2,230 conservation projects (1,390 in situ and 610 ex situ, 230 both) in 94 countries. They published 1,450 books, book chapters, journal articles, conference proceeding papers, posters and theses or dissertations. The publications can be searched using keywords, name of author, type of publication, institution name, or date.

### Regulation

In the United States, any public animal exhibit must be licensed and inspected by the United States Department of Agriculture (USDA), United States Environmental Protection Agency, Drug Enforcement Agency, Occupational Safety and Health Administration, and others. Depending on the animals they exhibit, the activities of zoos are regulated by laws including the Endangered Species Act, the Animal Welfare Act, the Migratory Bird Treaty Act of 1918 and others. Additionally, zoos in North America may choose to pursue accration by the Association of Zoos and Aquariums (AZA). The American association has developed a definition for zoological gardens and aquariums as part of its accration standards: “A permanent cultural institution which owns and maintains captive wild animals that represent more than a token collection and, under the direction of a professional staff, provides its collection with appropriate care and exhibits them in an aesthetic manner to the public on a regularly scheduled basis. They shall further be defined as having as their primary business the exhibition, conservation and preservation of the earth's fauna in an educational and scientific manner." To achieve accration, a zoo must pass an application and inspection process and meet or exceed the AZA's standards for animal health and welfare, fundraising, zoo staffing, and involvement in global conservation efforts. Inspection is performed by three experts (typically one veterinarian, one expert in animal care, and one expert in zoo management and operations) and then reviewed by a panel of twelve experts before accration is awarded. This accration process is repeated once every five years. The AZA estimates that there are approximately 2,400 animal exhibits operating under USDA license as of February 2007; fewer than 10% are accred.

### Airlift Filters

Sponge filters and corner filters (sometimes called box filters) work by essentially the same mechanism as an internal filter. Both generally work by airlift, using bubbles from an air pump rising in a tube to create flow. In a sponge filter, the inlet may only be covered by a simple open-cell block of foam. A corner filter is slightly more complex. These filters are often placed in the corner on the bottom of the aquarium. Water enters slits in the box, passes through a layer of medium, then exits through the airlift tube to return to the aquarium. These filters tend to only be suitable for small and lightly stocked aquaria. The sponge filter is especially useful for rearing fry where the sponge prevents the small fish from entering the filter.

### Undergravel Filters

One of the oldest types of filters, the undergravel filters consist of a porous plate which is placed beneath the gravel on the base of the aquarium and one, or more, uplift tubes. Historically, undergravel filters have been driven via air displacement. Air stones are placed at the base of uplift tubes which force water out of the uplift tube creating negative pressure beneath the undergravel filter plate. Water then percolates down through the gravel which itself is the filtration material. Greater flow rate of water through the gravel can be achieved via the use of water pump rather than air displacement.

Beneficial bacteria colonise the gravel bed and provide biological filtration, using the substrate of the aquarium itself as a biological filter.

Undergravel filters can be detrimental to the health of aquatic plants. Fine substrates such as sand or peat may clog an undergravel filter. Undergravel filters are not effective if the substrate bed is uneven. In an uneven gravel bed, water will flow only through the thin portions of the bed, leaving the more heavily covered areas to become anoxic. Because of this, animals that dig, such as cichlids, are best kept in an aquarium using some other type of filtration.

## Aquarium Furniture

Aquarium furniture refers to the various ornaments and functional items in an aquarium. Ornamental aquarium furniture is often kitsch: popular examples include ceramic mermaids, 'sunken' ships and castles, and the ever-popular (but curiously misplaced) "No Fishing" sign. Another strange piece of decor is the ubiquitous plastic corals found often in freshwater tanks.Examples of functional aquarium furniture would include devices for removing algae from the glass (either a razor or a scouring pad, attached to the glass by a magnet), airstones, water filters, water heaters, and food dispensers. Aquarium furniture may also refer to an item of (regular) furniture that features an aquarium in its design. A stand or cabinet that supports the aquarium may be considered aquarium furniture. Also, many home reef aquariums canopies containing metal halide lights. The canopies are often constructed to the same standards as high quality cabinetry.

## Coral

Corals are marine animals in class Anthozoa of phylum Cnidaria typically living in compact colonies of many identical individual "polyps". The group includes the important reefbuilders that inhabit tropical oceans and secrete calcium carbonate to form a hard skeleton. A coral "head" is a colony of myriad genetically identical polyps. Each polyp is a spineless animal typically only a few millimetres in diametre and a few centimetres in length. A set of tentacles surround a central mouth opening. An exoskeleton is excreted near the base. Over many generations, the colony thus creates a large skeleton that is characteristic of the species. Individual heads grow by asexual reproduction of polyps. Corals also breed sexually by spawning: polyps of the same species release gametes simultaneously over a period of one to several nights around a full moon.

Although corals can catch small fish and plankton, using stinging cells on their tentacles, most corals obtain the majority of their energy and nutrients from photosynthetic unicellular algae called zooxanthellae that live within the coral's tissue. Such corals require sunlight and grow in clear, shallow water, typically at depths shallower than 60 metres (200 ft).

Corals can be major contributors to the physical structure of the coral reefs that develop in tropical and subtropical waters, such as the enormous Great Barrier Reef off the coast of Queensland, Australia. Other corals do not have associated algae and can live in much deeper water, with the cold-water genus *Lophelia* surviving as deep as 3,000 metres (9,800 ft). Examples live on the Darwin Mounds located north-west of Cape Wrath, Scotland. Corals have also been found off the coast of the U.S. in Washington State and the Aleutian Islands in Alaska.

### Taxonomy

Corals divide into two subclasses, depending on the number of tentacles or lines of symmetry, and a series of orders corresponding to their exoskeleton, nematocyst type and mitochondrial genetic analysis. Common coral typing crosses suborder/class boundaries.

### Hermatypic Corals

Hermatypic corals in the subclass Scleractinia are stony corals that build reefs. They mostly obtain at least part of their energy requirements from zooxanthellae, symbioticphotosynthetic microalgae. They secrete calcium carbonate to form a hard skeleton. Those having six or fewer lines of symmetry in their body structure are called hexacorallia orZoantharia. This group includes reef-building corals (scleractinians), sea anemones and zoanthids. Hermatypic genera include *Scleractinia*, *Millepora*, *Tubipora* and *Heliopora*. In the Caribbean alone, at least 50 species of uniquely structured hard coral exist. Well-known types include:

- Brain corals grow to 1.8 metres (5.9 ft) in width.
- *Acropora* and staghorn corals grow fast and large, and are important reef-builders. Staghorn coral displays large, antler-like branches, and grows in areas with strong surf.
- Pillar coral forms pillars which can grow to 3 metres (9.8 ft) in height.
- *Leptopsommia*, or rock coral, appears almost everywhere in the Caribbean.

### Ahermatypic Corals

Ahermatypic corals have no zooxanthellae. They sport eight tentacles and are also called octocorallia. They include corals in subclass *Alcyonacea*, as well as some species in order *Anthipatharia* (black coral, *Cirripathes, Antipathes*). Ahermatypic corals, such as sea whips, sea feathers, and sea pens, are also known as soft corals. Unlike stony corals, they are flexible, undulating in the current, and often are perforated, with a lacy appearance. Their skeletons are proteinaceous, rather than calcareous. Soft corals are somewhat less plentiful (in the Caribbean, twenty species appear) than stony corals.

### Perforate Corals

Corals can be perforate or imperforate. Perforate corals have porous skeletons, which allows their polyps to connect with each other through the skeleton. Imperforate corals have hard solid skeletons.

### Anatomy

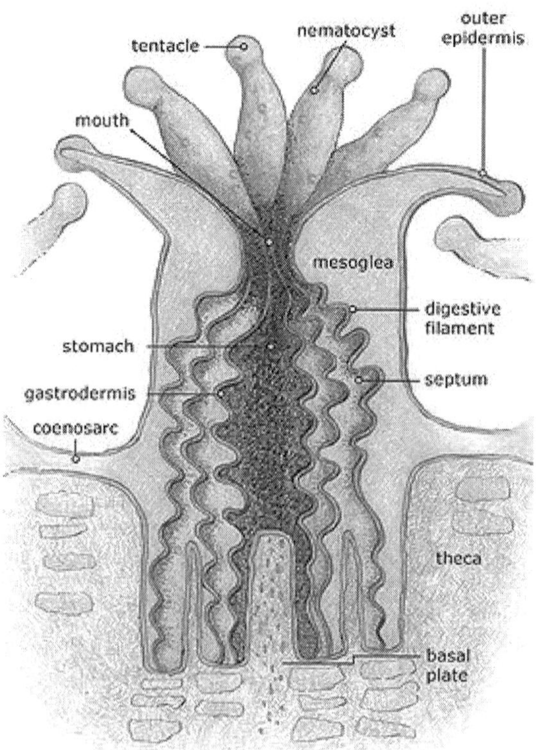


***Figure:*** *Anatomy of a coral polyp*

***Colonial Form:*** The polyps interconnect by a complex and well-developed system of gastrovascular canals, allowing significant sharing of nutrients and symbiotes. In soft corals, these range in size from 50–500 micrometres (0.0020–0.020 in) in diametre, and allow transport of both metabolites and cellular components.

***Polyp:*** While the coral head is the familiar visual form of a single organism, it is actually a group of many individual, yetgenetically identical, multicellular organisms known as polyps.

Polyps are usually a few millimetres in diametre, and are formed by a layer of outer epithelium and inner jellylike tissue known as the mesoglea.

They are radially symmetrical, with tentacles surrounding a central mouth, the only opening to the stomach or coelenteron, through which food is ingested and waste expelled.

***Exoskeleton:*** The stomach closes at the base of the polyp, where the epithelium produces an exoskeleton called the basal plate or calicle (L. small cup).

The calicle is formed by a thickened calcareous ring (annular thickening) with six supporting radial ridges.

These structures grow vertically and project into the base of the polyp.

When a polyp is physically stressed, its tentacles contract into the calyx so that virtually no part is exposed above the skeletal platform. This protects the organism from predators and the elements.

The polyp grows by extension of vertical calices which occasionally septate to form a new, higher, basal plate. Over many generations, this extension forms the large calcareous structures of corals and ultimately coral reefs.

Formation of the calcareous exoskeleton involves deposition of the mineral aragonite by the polyps from calcium and carbonate ions they acquire from seawater.

The rate of deposition, while varying greatly across species and environmental conditions, can reach 10 g/m$^2$ of polyp/day (0.3 ounce/sq yd/day).

This is light dependent, with night-time production 90% lower than that during the middle of the day.

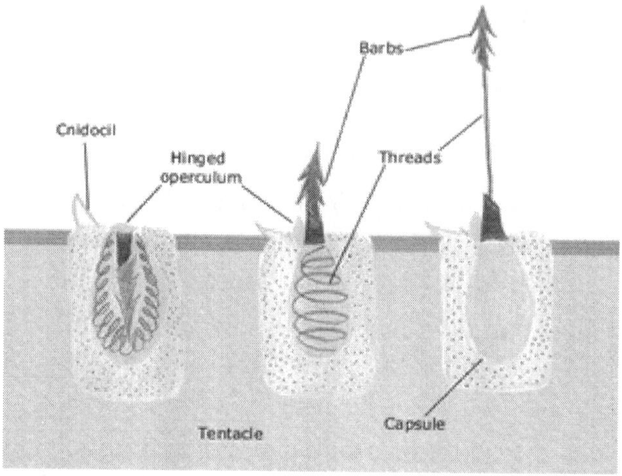


***Figure:*** *Nematocyst discharge: A dormant nematocyst discharges response to nearby prey touching the cnidocil, the operculum flap opens, and its stinging apparatus fires the barb into the prey, leaving a hollow filament through which poisons are injected to immobilise the prey, then the tentacles manoeuvre the prey to the mouth.*

***Tentacles:*** Nematocysts are stinging cells at the tips of the calices that carry venom, which they rapidly release in response to contact with another organism. The tentacles also bear a contractile band of epithelium called the pharynx. Jellyfishand sea anemones also carry nematocysts.

## Ecology

***Feeding:*** Polyps feed on a variety of small organisms, from microscopic plankton to small fish. The polyp's tentacles immobilise or kill prey using their nematocysts (also known as 'cnidocysts'). The tentacles then contract to bring the prey into the stomach. Once digested, the stomach reopens, allowing the elimination of waste products and the beginning of the next hunting cycle.

***Zooxanthellae Symbiote:*** Many corals, as well as other cnidarian groups such as Aiptasia (a sea anemone) form a symbiotic relationship with a class of algae, zooxanthellae, of the genus *Symbiodinium. Aiptasia*, a familiar pest among coral reef aquarium hobbyists, serves as a valuable model organism in the study of cnidarian-algal symbiosis. Typically, each polyp harbours one species of algae. Via photosynthesis, these provide energy for the coral, and aid in calcification. The algae benefit from a safe place to live and consume the polyp's carbon dioxide and nitrogenous waste. Due to the strain the algae can put on the polyp, stress on the coral often drives them to eject the algae. Mass ejections are known as coral bleaching, because the algae contribute to coral's brown colouration; other colours, however, are due to host coral pigments, such as green fluorescent proteins (GFPs). Ejection increases the polyp's chance of surviving short-term stress—they can regain algae, possibly of a different species at a later time. If the stressful conditions persist, the polyp eventually dies.

### Reproduction

Corals can be both gonochoristic (unisexual) and hermaphroditic, each of which can reproduce sexually and asexually. Reproduction also allows coral to settle in new areas.

### Sexual

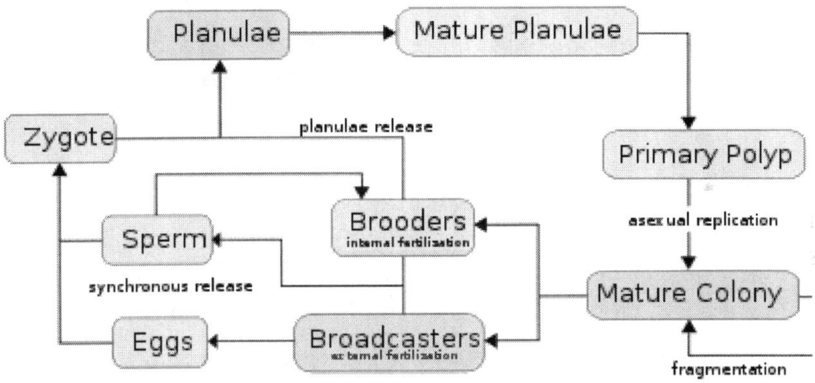


***Figure:*** *Life cycles of broadcasters and brooders*

Corals predominantly reproduce sexually. About 25% of hermatypic corals (stony corals) form single sex (gonochoristic) colonies, while the rest are hermaphroditic.

***Broadcasters:*** About 75% of all hermatypic corals "broadcast spawn" by releasing gametes—eggs and sperm—into the water to spread offspring. The gametes fuse during fertilization to form a microscopic larva called a planula, typically pink and elliptical in shape.

A typical coral colony forms several thousand larvae per year to overcome the odds against formation of a new colony. Synchronous spawning is very typical on the coral reef, and often, even when multiple species are present, all corals spawn on the same night. This synchrony is essential so male and female gametes can meet. Corals rely on environmental cues, varying from species to species, to determine the proper time to release gametes into the water.

The cues involve temperature change, lunar cycle, day length, and possibly chemical signalling. Synchronous spawning may form hybrids and is perhaps involved in coral speciation. The immediate cue is most often sunset, which cues the release. The spawning event can be visually dramatic, clouding the usually clear water with gametes.

***Brooders:*** Brooding species are most often ahermatypic (not reef-building) in areas of high current or wave action. Brooders release only sperm, which is negatively buoyant, sinking on to the waiting egg carriers who harbour unfertilized eggs for weeks. Synchronous spawning events sometimes occurs even with these species. After fertilization, the corals release planula that are ready to settle.

***Planulae:*** Planulae exhibit positive phototaxis, swimming towards light to reach surface waters, where they drift and grow before descending to seek a hard surface to which they can attach and begin a new colony.

They also exhibit positive sonotaxis, moving towards sounds that emanate from the reef and away from open water. High failure rates afflict many stages of this process, and even though millions of gametes are released by each colony, few new colonies form.

The time from spawning to settling is usually two to three days, but can be up to two months. The larva grows into a polyp and eventually becomes a coral head by asexual budding and growth.

### *Asexual*

Within a coral head, the genetically identical polyps reproduce asexually, either via gemmation (budding) or by longitudinal or transversal division, both shown in the photo of*Orbicella annularis*. Budding involves splitting a smaller polyp from an adult. As the new polyp grows, it forms its body parts. The distance between the new and adult polyps grows, and with it, the coenosarc.

Budding can be:

- Intratentacular—from its oral discs, producing same-sized polyps within the ring of tentacles
- Extratentacular—from its base, producing a smaller polyp

***Figure:*** *Calices (basal plates) of* Orbicella annularis *showing multiplication by gemmation (small central calice) and division (large double calice)*

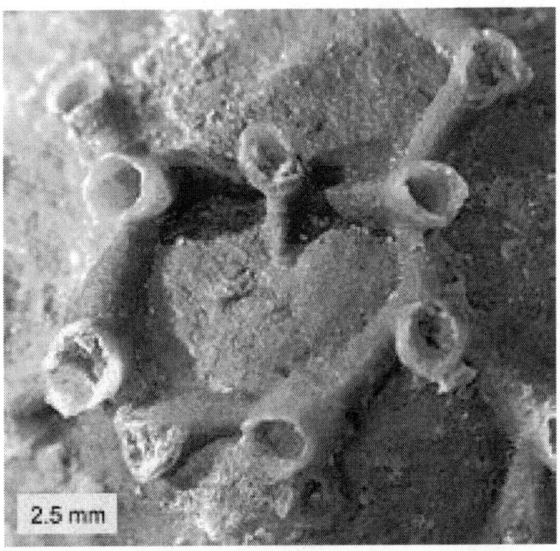


***Figure:*** *The tabulate coral* Aulopora *(Devonian) showing initial budding from protocorallite*

Division forms two polyps each as large as the original. Longitudinal division begins when a polyp broadens and then divides its coelenteron, analogous to splitting a log along its length. The mouth also divides and new tentacles form. The two "new" polyps then generate their missing body parts and exoskeleton. Transversal division occurs when polyps and the exoskeleton divide transversally into two parts. This means one has the basal disc (bottom) and the other has the oral disc (top), similar to cutting the end off a log. The new polyps must separately generate the missing pieces. Asexual reproduction has several benefits for these sessile colonial organisms:

- Cloning allows high reproduction rates, supporting rapid habitat exploitation.
- Modular growth allows biomass to increase without a corresponding decrease in surface-to-volume ratio.
- Modular growth delays senescence, by allowing the clone-type to survive the loss of one or more modules.
- New modules can replace dead modules, reducing clone-type mortality and preserving the colony's territory.
- Spreading the clone type to distant locations reduces clone-type mortality from localised threats.

***Colony Division:*** Whole colonies can reproduce asexually, forming two colonies with the same genotype.

- Fission occurs in some corals, especially among the family Fungiidae, where the colony splits into two or more colonies during early developmental stages.
- Bailout occurs when a single polyp abandons the colony and settles on a different substrate to create a new colony.
- Fragmentation involves individuals broken from the colony during storms or other disruptions. The separated individuals can start new colonies.

## Reefs

Evolutionary history

***Figure:*** *Solitary rugose coral (*Grewingkia*) in three views; Ordovician, southeastern Indiana*

Although corals first appeared in the Cambrian period, some 542 million years ago, fossils are extremely rare until the Ordovician period, 100 million years later, when rugose and tabulate corals became widespread.

Tabulate corals occur in limestones and calcareous shales of the Ordovician and Silurian periods, and often form low cushions or branching masses alongside rugose corals. Their numbers began to decline during the middle of the Silurian period, and they became extinct at the end of the Permian period,250 million years ago. The skeletons of tabulate corals are composed of a form of calcium carbonate known as calcite.Rugose corals became dominant by the middle of the Silurian period, and became extinct early in the Triassic period. The rugose corals existed in solitary and colonial forms, and were also composed of calcite.

The scleractinian corals filled the niche vacated by the extinct rugose and tabulate species. Their fossils may be found in small numbers in rocks from the Triassic period, and became common in the Jurassic and later periods. Scleractinian skeletons are composed of a form of calcium carbonate known as aragonite. Although they are geologically younger than the tabulate and rugose corals, their aragonitic skeleton is less readily preserved, and their fossil record is less complete.At certain times in the geological past, corals were very abundant. Like modern corals, these ancestors built reefs, some of which ended as great structures in sedimentary rocks.

Fossils of fellow reef-dwellers algae, sponges, and the remains of many echinoids, brachiopods, bivalves, gastropods, and trilobites appear along with coral fossils. This makes some corals useful index fossils that enabled geologists to date the rocks in which they are found. Coral fossils are not restricted to reef remnants, and many solitary fossils may be found elsewhere, such as *Cyclocyathus*, which occurs in England's Gault clay formation.

A Petoskey stone is a rock and a fossil, often pebble-shaped, that is composed of a fossilized coral, *Hexagonaria percarinata*. They are found predominantly in Michigan's Upper Peninsula, and the northwestern portion of Michigan's Lower Peninsula.

***Threats:*** Coral reefs are under stress around the world. In particular, coral mining, agricultural and urban runoff, pollution (organic and inorganic), overfishing, blast fishing, disease, and the digging of canals and access into islands and bays are localised threats to coral ecosystems. Broader threats are sea temperature rise, sea level rise and pH changes from ocean acidification, all associated with greenhouse gas emissions. In 1998, 16% of the world's reefs died as a result of increased water temperature.

General estimates show approximately 10% of the world's coral reefs are dead. About 60% of the world's reefs are at risk due to human-related activities. The threat to reef health is particularly strong in Southeast Asia, where 80% of reefs are endangered. Over 50% of the world's coral reefs may be destroyed by 2030; as a result, most nations protect them through environmental laws. In the Caribbean and tropical Pacific, direct contact between ~40 to 70% of common seaweeds and coral causes bleaching and death to the coral via transfer of lipid–solublemetabolites. Seaweed and algae proliferate given adequate nutrients and limited grazing by herbivores such as parrotfish.

Water temperature changes of more than 1-2 degrees Celsius (1.8-3.6 degrees Fahrenheit) or salinity changes can kill coral. Under such environmental stresses, corals expel their zooxanthellae; without them coral tissues reveal the white of their skeletons, an event known as coral bleaching. Submarine springs found along the coast of Mexico's Yucatán Peninsula produce water with a naturally low pH (a measure of acidity) providing conditions similar to those expected to become widespread as the oceans absorb carbon dioxide. Surveys discovered multiple species of live coral that appeared to tolerate the acidity. The colonies were small and patchily distributed, and had not formed structurally complex reefs such as those that compose the nearby Mesoamerican Barrier Reef System.

***Protection:*** Many governments now prohibit removal of coral from reefs, and inform coastal residents about reef protection and ecology. While local action such as habitat restoration and herbivore protection can reduce local damage, the longer-term threats of acidification, temperature change and sea-level rise remain a challenge. To eliminate destruction of corals in their indigenous regions, projects have been started to grow corals in non-tropical countries.

### Relation to Humans

Local economies near major coral reefs benefit from an abundance of fish and other marine creatures as a food source. Reefs also provide recreational scuba diving and snorkeling tourism. These activities can damage coral.

In medicine chemical compounds from corals are used for cancer, AIDS, pain, and other uses. Coral skeletons, e.g., *Isididae* are also used for bone grafting in humans.

Live coral is highly sought after for aquaria. Soft corals are easier to maintain in captivity than hard corals.

***Jewellery:*** Coral's many colours give it appeal for necklaces and other jewellery. Intensely red coral is prized as a gemstone. Sometimes called fire coral, it is not the same as fire coral. Red coral is very rare because of overharvesting.

***Construction:*** Coral reefs on land provide lime for use as building blocks ("coral rag"). Coral *rag* is an important local building material in places such as the East African coast.

***Climate Research:*** The annual growth bands in deep sea bamboo corals (*Isididae*) and others may be among the ocean's first organisms to display the effects of ocean acidification. They produce growth rings similar to those of trees, and can provide a view of changes in the condition in the deep sea over time. They allow geologists to construct year-by-year chronologies, a form of incremental dating, which underlie high-resolution records of past climatic and environmental changes using geochemical techniques. Certain species form communities called microatolls, which are colonies whose top is dead and mostly above the water line, but whose perimetre is mostly submerged and alive. Average tide level limits their height.

By analysing the various growth morphologies, microatolls offer a low resolution record of sea level change. Fossilized microatolls can also be dated using radioactive carbon dating. Such methods can help to reconstruct Holocene sea levels.

***Aquaculture:*** Coral aquaculture, also known as *coral farming* or *coral gardening*, is the cultivation of corals for commercial purposes or coral reef restoration. Aquaculture is showing promise as a potentially effective tool for restoring coral reefs, which have been declining around the world. The process bypasses the early growth stages of corals when they are most at risk of dying. Coral seeds are grown in nurseries then replanted on the reef. Coral is farmed by coral farmers who live locally to the reefs and farm for reef conservation or for income. It is also farmed by scientists for research, by businesses for the supply of the live and ornamental coral trade and by private aquarium hobbyists.

## Reef Aquarium

***Figure:*** *Reef aquarium in Monaco*

A reef aquarium or reef tank is a marine aquarium that prominently displays live corals and other marine invertebrates as well as fish that play a role in maintaining the tropical coral reef environment. A reef aquarium requires appropriately intense lighting, turbulent water movement, and more stable water chemistry than fish-only marine aquaria, and careful consideration is given to which reef animals are appropriate and compatible with each other.

### Components

Reef aquariums consist of a number of components, in addition to the livestock, including:

*Display tank:* The primary tank in which the livestock are kept and shown.

*Stand:* A stand allows for placement of the display tank at eye level and provides space for storage of the accessory components.

*Sump:* An accessory tank in which mechanical equipment is kept. A remote sump allows for a clutter-free display tank.

*Refugium:* An accessory tank dedicated to the cultivation of beneficial macroalgae and microflora/fauna. The refugium and sump are often housed in a single tank with a system of dividers to separate the compartments.

*Lighting:* Several lighting options are available for the reef-keeper and are tailored to the types of coral kept.

*Canopy:* The canopy houses the light fixtures and provides access to the tank for feeding and maintenance.

*Filtration and water movement:* A variety of filtration and water movement strategies are employed in reef aquaria. Bulky equipment is often relegated to the sump.

### Display Tank

***Aquarium:*** Quascaping, Sump (aquarium), and Refugium (fishkeeping)

fig 1

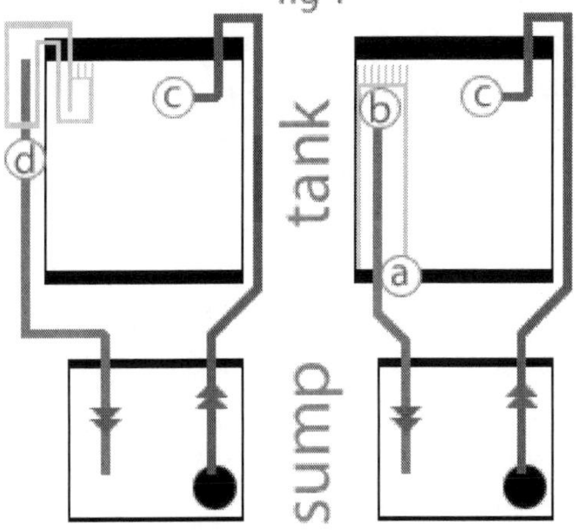


A "reef ready" or simply "drilled" tank is often used. This style of tank has holes drilled into the rear pane allowing water to drain into the sump or refugium. These drains are usually housed in an internal overflow apparatus made of plastic or glass which encloses a drain standpipe and a water return line. The surface water pours over the overflow, down the standpipe, through PVC piping, into the sump. After transiting the sump, water is pushed by a return water pump through the second hole and into the aquarium. Alternatively, standard non-drilled aquariums employ an external "hang-on" overflow that feeds water via continuous siphonto the sump. The tanks are usually constructed from either glass or acrylic. Acrylic has the advantage of optical clarity, lightness, and ease of drilling. Drawbacks include a tendency to scratch easily, bowing, and often limited access from above due to top bracing. Glass aquariums are heavier but harder to scratch. Other materials such as epoxy-coated plywood have been used by industrious DIYers.

### Filtration

The primary biological filtration for reef aquariums usually comes from the use of live rock which come from various tropical zones around existing reefs, or more recently aquacultured rock from Florida. Some reef keepers also use what is called deep sand beds (DSB). These are often employed to augment the biological filtration by aiding in the reduction of nitrate, a waste product in an incomplete nitrogen cycle. Deep sand bed opponents may prefer a "bare bottom" or "suspended reef" which allows for easier removal of nitrate-generating accumulated detritus. This biological filtration is usually supplemented by protein skimmers. Protein skimmers use the foam fractionation process wherein air is introduced into a water stream creating microbubbles. Organic waste adheres to the surface of these microbubbles and is removed as it overflows at the reactor surface into a removable cup. This group of elements used in conjunction is characteristic of the Berlin Method, named for the city in which it was first devised.

In recent years the Berlin Method is often supplemented with a refugium. A refugium provides many benefits, which include nitrate reduction, as well as providing a natural food source. It typically houses two main species of macroalgae, including *Caulerpa prolifera* or *chaetomorphae* or both (because these two strains are known to not spore but grow by rooting to propagate). Macroalgae is used for two reasons: to remove from the water excess nutrients such as nitrate, phosphate, and iron, and to support beneficial microflora and fauna (zooplankton). Small invertebrates (copepods and amphipods) are provided a space free of predation to grow and, when returned to the display tank, serve as food for corals and fish. Conventional combined mechanical/biological filtration used in fish only systems is avoided because those filters trap detritus and produce nitrate which may stunt or even kill many delicate corals. Chemical filtration in the form of activated carbon is used when needed to remove discolouration of the water, or to remove dissolved matter (organic or otherwise) to help purify the water in the reef system.

### Water Movement

Water movement is important in the reef aquarium with different types of coral requiring different flow rates. At present, many hobbyists advocate a water turnover rate of 10x: 10 x aquarium capacity in gallons = required flow in gallons per hour. This is a general rule with many exceptions. For instance, Mushroom Coral requires little flow and is commonly found in crevices near the base of the reef. Species such as Acropora and Montipora thrive under much more turbulent conditions in the range of 30 to 40 times more flow, which imitates breaking waves in shallow water near the tip of the reef. The directions which water pumps are pointed within an aquarium will have a large effect on flow speeds. Many corals will gradually move themselves to a different area of the tank if the water movement in its current area is not satisfactory.

"Since flow speed is the critical measure for determining the rate of gas exchange, turnover does little to convey how fast a coral will respire and photosynthesize."

Reef ready tanks obtain at least a portion of the required water motion from the pump that returns water from the sump. This flow usually is augmented by other strategies. A popular strategy is placement within the display tank of multiple powerheads. Powerheads are simply small submersible water pumps that produce a laminar or narrow, unidirectional water stream. If the presence of the powerhead in the tank does not fit with the aesthetics of the display, small holes may be drilled in an overflow of a tank and the bulk of the powerhead can be hidden, leaving only the small funnel spout visible in the tank. The pumps may be alternately switched on and off using a wave timer and aimed at one another or at the aquarium glass to create turbulent flow in the tank. Drawbacks to the use of these powerheads include their capacity to clutter the display tank, propensity for excess heat production, and the laminar quality of water flow often produced. Another method is the closed loop in which water is pulled from the main tank into a pump which returns the water back into the aquarium via one or more returns to create water turbulence. Newer submersible propeller pumps are gaining popularity and are able to generate large volumes of turbulent water flow without the intensely directed laminar force of a power head. Propeller pumps are more energy-efficient than powerheads, but require a higher initial investment.

Another recent method is the gyre tank. A gyre tank encourages a maximum amount of water momentum through a divider in the centre of the aquarium. The divider leaves an open, unobstructed space which provides a region with little friction against water movement. Building water momentum using a gyre is an efficient method to increase flow, thus benefiting coral respiration and photosynthesis. Water flow is important to bring food to corals, since no coral fully relies on photosynthesis for food. Gas exchange occurs as water flows over a coral, bringing oxygen and removing gases and shedding material. Water flow assists in reducing the risk of thermal shock and damage by reducing the coral's surface temperature. The surface temperature of a coral living near the water's surface can be significantly higher than the surrounding water due to infrared radiation.

### *Lighting*

With the advent of newer and better technologies, increasing intensities and a growing spectrum, there are many options to consider.

Many, if not most aquarium corals contain within their tissue the symbiotic algae called zooxanthellae. It is these zooxanthellae that require light to perform photosynthesis and in turn produce simple sugars that the corals utilise for food. The challenge for the hobbyist is to provide enough light to allow photosynthesis to maintain a thriving population of zooxanthellae in a coral tissue. Though this may seem simple enough, in reality this can prove to be a very complex task.

Some corals such as mushroom corals and polyp corals require very little light to thrive. Conversely, large-polyp stony corals such as brain coral, bubble coral, elegance coral, cup coral, torch coral, and trumpet coral require moderate amounts of light, and small polyp stony corals such as *Acropora*, *Montipora*, *Porites*, and *Pocillopora* require high intensity lighting.

Of the various types, most popular aquarium lighting comes from metal halide lamps, very high output or VHO, compact fluorescent and T5 high output lighting systems. Although they were once widely used, many reef tank aquarists have abandoned T12 and T8 fluorescent lamps due to their poor intensity, and mercury vapour due to its production of a limited light spectrum.

Recent advances in lighting technology have also made available a completely new technology for aquarium lighting: lightemitting diodes (LEDs). Although LEDs themselves are not new, the technology has only recently been adapted to produce systems with qualities that allow them to be considered viable alternatives to gas- and filament-based aquarium lighting systems. The newness of the technology does cause them to be relatively expensive, but these systems bring several advantages over traditional lighting. Although their initial cost is much higher, they tend to be economical in the long run because they consume less power and have far longer lifespans than other systems. Also, because LED systems are made of hundreds of very small bulbs, a microcomputer can control their output can be controlled to simulate daybreak and sunset. Some systems also have the ability to simulate moonlight and the phases of the moon, as well as vary the colour temperature of the light produced. The choices for aquarium lighting are made complicated by variables such as colour temperature, (measured in kelvins), colour rendering index (CRI), photosynthetically active radiation (PAR) and lumens. Power output available to the hobbyist can range from a meager 9 W fluorescent lamp to a blinding 1000 W metal halide lamp. Lighting systems also vary in the light output produced by each bulb type—listed in order of weakest to strongest they would be: T8/12 or normal output lamps, compact fluorescent and T5 high output, VHO, and metal halide lamps. To further complicate matters, there are several types of ballasts available: electric ballast, magnetic ballast, and pulse start ballast.

### Heating and Cooling

Reef tanks are usually kept at a temperature between 25 and 28 °C (75-82 °F). Radical temperature shifts should be avoided as these can be particularly harmful to reef invertebrates and fish. Depending on the location of the tank and the conditions therein (i.e. heat/air conditioning), one may install a heater and/or a chiller for the tank. Heaters are relatively inexpensive and readily available at any local fish store. Aquarists frequently use the sump to hide unsightly equipment such as heaters. Chillers, on the other hand, are expensive and are more difficult to locate. For many aquarists, installing surface fans and running home air conditioning suffice in place of a chiller. Fans cool the tank via evapourative cooling and require more frequent top-off of the aquarium water.

### *Water Chemistry*

Stony corals, which are defined by their calcerous calcium carbonate skeletons ($CaCO_3$), are the focus of many advanced reef keepers. These corals require additional attention to water chemistry, especially maintenance of stable and optimal calcium, carbonate, and pH levels.

These parameters may be tracked and adjusted with test kits and frequent manual dosing of calcium and pH buffer additives requiring no additional equipment. Alternatively, automated methods employing small dedicated computers with electronic water quality monitoring capabilities are often used to control water chemistry parameters via several components including calcium reactors and kalkwasser reactors. Calcium reactors are canisters filled with crushed coral skeletons. Carbon dioxide is injected into the canister acidifying the water and dissolving the coral skeletons. The acidified and $CaCO_3$ rich solution is then pumped into the sump. The excess $CO_2$ then diffuses out of the water and into the air leaving behind the $CaCO_3$. Kalkwasser is an aqueous solution of calcium hydroxide, $Ca(OH)_2$. The kalk reactor stirs the solution, preventing precipitation, and dispenses the solution into the sump where the $Ca(OH)_2$ combines with dissolved $CO_2$ to produce $CaCO_3$. These components must be controlled by a computer to prevent dangerous changes in pH due to the acidic calcium reactor effluent or alkaline kalkwasser effluent.

Optimal water parameters are:

- Salinity: 1.022–1.025 sg or 30–34 parts per thousand
- Temperature: 76–80° Fahrenheit
- Ammonia ($NH_3$): 0 ppm (parts per million)
- Nitrite ($NO_2^-$): 0 ppm
- Nitrate ($NO_3^-$): 0–10 ppm
- Phosphate ($PO_4^{-3}$): 0 ppm
- pH: 8.2–8.6
- Calcium ($Ca^{2+}$): 400–450 ppm
- Alkalinity: 7–12 dKH

Trace elements can become depleted by marine livestock and filtration, and can be replenished during a water change.

### Safety

Large volumes of highly conductive salt water, complex plumbing, and numerous electrical appliances housed in close proximity certainly pose a significant risk of damage to both person and property and require close attention to safety. All equipment should be used according to manufacturer instructions. Electrical equipment should be placed above water level whenever possible, and drip loops should always be used. Circuit limits should never be exceeded and all appliances should be plugged into ground fault circuit interrupter (GFCI) outlets. These can be purchased at any hardware store and are relatively easy to install. Plug in GFCI power strips are also readily available. Home monitoring equipment with water sensors can also be adapted for the home aquarist and used to alert the owner of power outages or water overflows. This equipment can allow for timely intervention in a potential disaster and provides an added sense of security for frequent travellers.

### Nano Reefs

A nano reef is a type of marine aquarium that is typically less than 140 litres (30 gallons). The exact limit that distinguishes a nano reef from a regular reef is somewhat ill-defined (some claim that anything less than 180 litres/40 gallons would qualify), but 140 litres (30 gallons) seems to be the generally accepted limit. Nano reefs have become quite popular in recent years among fish keeping hobbyists, primarily because of their smaller size, maintainability, and the possibility of lower costs. The burgeoning interest in this niche of marine aquarium science has fostered several notable contributions ranging from specific consumer products such as specialised aquarium filters, compact high intensity lighting systems and smaller circulation pumps. Such equipment allows the aquarist to maintain an environment wherein many marine organisms are capable of thriving. Nano reefs are very commonly sold as complete kits which contain the tank, stand, power compact T5, T8, PL lamps or metal halide lighting, protein skimmer, UV steriliser, 3 or more stage filtration, a heater and a water pump or powerhead. However, many nano reef keepers decide to upgrade their aquariums with better quality equipment such as a more powerful protein skimmer or lighting.

### Pico Reefs

Another term gaining popularity is *pico reef*, which is used to refer to the smallest of nano reef aquariums. Most online forum polls set the range of approximately 5.5 gallons and below as pico reefs. These tiny tanks require even more diligence with regard to water changes and attention to water chemistry because the small water volume provides little room for error.

Care must be exercised when stocking these tiny tanks because too many inhabitants can easily overload the tank's ability to process wastes effectively. For the smallest of pico reefs, even the presence of a single fish is discouraged. Pico reefs often consist of live rock, hardy corals, and small invertebrates such as hermit crabs and marine snails.

The keeping of pico reef aquariums has tested the extent of allelopathy, the chemical and physical means by which corals compete for space. Before the advent of these concentrated environments, it was thought to be impossible for corals of even a few mixed genera to occupy such a small shared water volume.

### Challenges Associated with Small Reef Aquariums

Because of the small water volume, nano reef aquariums require extra attention to water quality compared to aquariums of larger water volumes. Many experienced reef aquarists recommend testing the water twice weekly, with water changes at least every week.

In particular, ammonia, nitrite, nitrate, pH, salinity, alkalinity, calcium and phosphate levels should be monitored closely. When it comes to nano reefs, even minute changes in water conditions such as mild temperature fluctuations can be problematic, whereas the greater water volume of larger aquariums provides a more stable and flexible environment.

Nano reefs also require extra care in the selection of occupants. There are two major factors to be considered: biological load, i.e. the ability of the tank to process the wastes produced by the occupants, and species compatibility. These issues, though present in larger tanks, are magnified in the nano tank. Species considered reef safe and able to coexist in larger tanks may not do well in a nano tank due to their close physical proximity. For this reason, smaller species of fish such as gobies and clownfish are popular choices due to their relatively small size and ability to coexist peacefully with other tank inhabitants.

### *Filtration in Nano Reefs*

Many nano reef aquarists prefer their displays to be as natural-looking as possible, and therefore choose to use as few mechanical filtration methods as possible. A primary filtration method in nano reefs is live rock and live sand, which are pieces of rock and sand that have broken from the coral reef and are populated with beneficial bacteria and other organisms that aid in breaking down organic wastes produced by larger organisms in the nano reef. Other nano reef aquarists use devices such as protein skimmers to remove excess waste from the aquarium, before it has a chance to be broken down to nitrate. Removing the excess wastes mechanically can reduce the frequency of water changes needed to keep nitrate levels low. Delaying the action of the mechanical filters, such as by means of a day-night timer, can allow invertebrates to filter-feed naturally. A refugium may also be used to export nutrients, when packed with macroalgae such as *Chaetomorpha*, and live rock. Deep sand bed filters are another filtration method.

Recently there have been several "natural" methods of processing waste in the aquarium and specifically small environments as nano-reefs. The research on the encouragement of the development of different types of sponges and micro-organisms to process the pollutants in the aquarium, a matter that has been gaining popularity in the aquarium community.

### *Freshwater*

Freshwater fishkeeping is by far the most popular branch of the hobby, with even small pet stores often selling a variety of freshwater fish, such asgoldfish, guppies, and angelfish. While most freshwater aquaria are community tanks containing a variety of compatible species, single-species breeding aquaria are also popular. Livebearing fish such as mollies and guppies are among those most easily raised in captivity, but aquarists also regularly breed many types of cichlid, catfish, characin, and killifish.

Many fishkeepers create freshwater aquascapes where the focus is on aquatic plants as well as fish. These aquaria include "Dutch Aquaria", named for European aquarists who designed them. In recent years, one of the most active advocates of the heavily planted aquarium is the Japanese aquarist Takashi Amano.

Garden ponds are in some ways similar to freshwater aquaria, but are usually much larger and exposed to ambient weather. In the tropics, tropical fish can be kept in garden ponds, but in the temperate zone species such as goldfish, koi, and orfe work better.

### *Saltwater*

Marine aquaria are generally more difficult to maintain and the livestock is significantly more expensive. As a result this branch tends to attract more experienced fishkeepers. Marine aquaria can be exceedingly beautiful, due to the attractive colours and shapes of the corals and the coral reef fish they host. Temperate zone marine fish are not as commonly kept in home aquaria, primarily because they do not thrive at room temperature. Coldwater aquaria must provide cooler temperature via a cool room (such as an unheated basement) or a refrigeration device known as a 'chiller'.

Marine aquarists often attempt to recreate a coral reef in their aquaria using large quantities of living rock, porous calcareous rocks encrusted with coralline algae, sponges, worms, and other small marine organisms. Larger corals as well as shrimps, crabs, echinoderms, and mollusks are added later on, once the aquarium has matured, as well as a variety of small fish. Such aquaria are sometimes called reef tanks.

# 5

# Elective Breeding

Selective breeding proved to be the most effective process for genetic improvement of plants and animals. Achievements in increased yields of paddy, wheat, milk, meat etc. during recent decades are due in large to selection. In aquaculture, the potential has not been fully exploited through selection. However, the selection work on Atlantic salmon in Norway, the Channel catfish in the USA and the Nile tilapia in the Philippines is enough testimony to advocate that it is quite possible to improve many economic traits like, survival, growth, resistance to disease, meat quality etc. through selection. In fact, many hatchery breeding programs result in lower quality fish through inbreeding depression.

In most of the developing countries, particularly in India and other South-east Asian countries, a substantial number of people depend on fishing or fish culture for their living. Consequently, there is over-exploitation due to heavy fishing both in marine and riverine areas reducing the size of populations in the absence of proper replenishment of the depleted stocks.

In aquaculture ever since the success of induced breeding of the Indian major carps through hypophysation, the technique has been gradually popularised throughout the Indian subcontinent as well as neighbouring countries such as Pakistan, Bangladesh, Burma, Nepal etc. Consequently, many hatcheries have developed in almost all these countries. However, in over four decades, attention has been paid only to produce fish seed to meet a pre-targeted quantity with very little regard to quality.

A survey of carp hatcheries in the state of Karnatak in South India was made (Eknath and Doyle 1990) because of the growing concern that hatchery stocks may be inbred, as suggested by rapid deterioration of their performance.

Eknath and Doyle (1990) found that the rate of the accumulation of inbreeding (F) was particularly high for the three most desirable carp species in the country, catla, rohu and mrigal.

The most important reasons for this appeared to be that all the hatcheries function as isolated genetically closed units, raising their own stocks of brood fish.

Moreover, the individuals for replenishment of brood stock in the hatchery are those which were retained after the hatchery had reached its seed production target for a given year (Eknath and Doyle 1990).

It was also reported that small hatcheries appear to rapidly inbreed their stocks.

The tendency to breed slow growing and later maturing individuals was also observed in some of the hatcheries, which is undersirable.

The survey Basavaraju *et al.* (1997) conducted particularly for catla in the three major seed producing hatcheries in the state of Karnataka, namely Tunga Bhadra Dam (TBD), Bhadra Reservoir Project (BRP) and Kabiri Reservoir Project (KRP), also indicated that the practices followed by these hatcheries are likely to result in both inbreeding and negative selection.

Maheshwari and Biradar (1998) reported decline in fertility of hatchery bred *Labeo rohita* due to inbreeding.

A study was conducted at Powarkheda, Freshwater Fish Farm of the Central Institute of Fisheries Education (ICAR), India where induced breeding of *L. rohita* was carried out between 1982 and 1992 to assess fertility.

It has been reported that breeding of related individuals up to the sixth generation led to 71% decline in fertility level as compared to base population.

It was also observed that the range, standard deviation and coefficient of variation were highest for fecundity in the first generation. This indicates, according to the authors, the scope for improvement through selection.

## Selection Work on Indian Major Carps in India

Currently two projects are in operation on carp selection. One on *Labeo rohita* as already mentioned, the other on *Catla catla* at the University of Agricultural Sciences, Bangalore, Karnataka State.

### Selective Breeding of Rohu

Founder populations of *L. rohita* have been collected from five different rivers in North India, namely, Rivers Ganga, Gomati, Yamuna, Sutlej and Brahmaputra, which are the native habitat of rohu. Rohu stock from the fish farm of CIFA has also been added to the gene pool. Thus the base population for the selection project consisted of rohu, from six sources, to broaden the gene pool.

All these stocks were collected from fry to fingerling stages and reared separately in earthen ponds for a period of one to two months before they were physically marked and stocked in communal ponds for raising to brood fish.

Marking was done using dye, fin clipping and a combination of both. Ganga rohu were dye injected with M-procian blue at the base of the right pectoral fin and Yamuna rohu at the base of the left pectoral fin.

In the case of Gomati and farm rohu the right and left pelvic fins were clipped respectively, while the Sutlej rohu were marked with the blue dye, inject at the base of the left pectoral fin by clipping the right pelvic fin. The reverse has been done to mark the Brahmaputra rohu, i.e. by injecting blue at the base of the right pectoral and clipping the left pelvic fin (Gupta, 1994).

During 1993, the first groups of full sibs were produced from the local and Ganga stocks as these two stocks were matured. Two or three males were mated with one female (paternal half sibs).

However, during the year 1994, full-sib groups (families) were produced from all the six stocks of rohu: two or three females were mated to a male, producing maternal half-sibs.

Rearing full-sib groups to tagable size fingerlings has been done in earthen nursery ponds of 100 $m^2$ for four to five months (Saha *et al.*, 1997). After the full-sib attained suitable tagging size (8–10 g) the fish were tagged by using Passive Integrated Transponder (PIT) tags, before they are stocked in ponds for communal rearing under both mono and polyculture systems. (Mahapatra *et al.*, 1998).

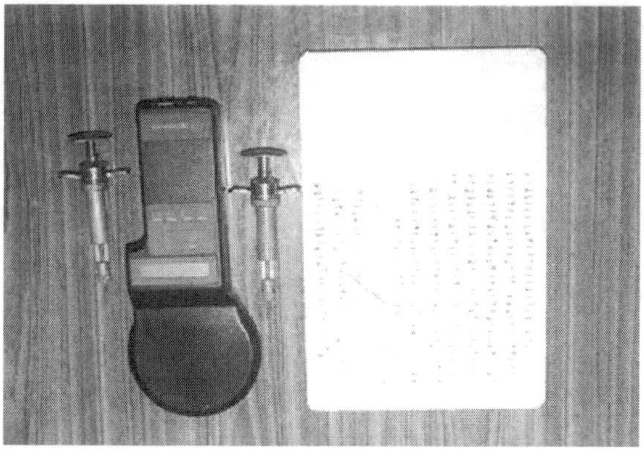

***Figure :** Passive integrated transponder (PIT) tags, tag implanter and tag reader*

After a period of one year, data pertaining to growth (length and weight) was collected and analysed to evaluate the individual breeding value and to rank the fish. The data so far collected indicated that growth and survival of the wild rohu stocks were equal to or better than the farmed stock.

This indicated that the present procedures followed by hatcheries for seed production are improper. A substantial additive genetic variation has been observed for growth rate and survival of rohu, both under mono and polyculture systems, thus indicating that growth rate and survival of rohu can be improved through selective breeding.

Competition for food may prohibit rohu from exhibiting its growth potential in polyculture. Proper feeding and pond management is essential while testing fish in a breeding programme.

The study so far has also indicated that stocks and full-sib groups within stocks ranked consistently high for growth performance in mono and polyculture, which suggests that there is no need for the development of special rohu lines for different systems. However, further studies are envisaged to test genotype by environment interaction, i.e. under intensive/extensive production systems and in different climatic zones.

## Selective Breeding of Catla

The main objectives of this project are to determine the extent of the problem of inbreeding in domesticated stocks of *Catla catla* in Karnataka state and to design a breeding programme for the improvement of the cultured catla stocks.

To this extent the methodologies being used are survey and data collection from state hatcheries for the assessment of the genetic status of the catla stocks presently being used for seed production and culture. The hypothesis of whether the domesticated stocks are inbred due to poor broodstock management will be tested by comparing the level of genetic variation between these and wild stock of catla obtained from rivers, Ganga, Brahmaputra and Mahanadi and by allozyme, mitochondrial DNA (RAPD) and tetranucleotide microsatillite electrophoretic techniques.

Protocols are planned for improving the genetic management of domesticated carp stocks in the state and to disseminate through on-station and on-farm comparative performance testing and by the use of best strains for selective breeding to further enhance their commercial importance.

### Progress to Date

A survey of the three main catla producing hatcheries of the state, TBD, BRP and KRP have indicated that the common practices of brood stock replacement seems to be to retain a few fingerlings each year which are normally collected towards the end of the spawning season, consisting of not more than two or three families.

These practices are likely to result in inbreeding and selection for poor growth (Basavaraju *et al.*, 1997). Genetic analysis of the hatchery stocks using four tetranucleotide microsatellite loci has shown these stocks to be different from each other.

Results of growth trials indicated no significant differences in the performance between the stocks of the key hatcheries. However, the interhatchery crosses produced highly significant increases in yield.

The rate of in breeding in smaller hatcheries was found to be unacceptably high which may have brought about a reduction in the culture performance of the hatchery bred seed.

Basavaraju *et al.* (1997) also reported the performance of intra and inter strain crosses between individuals of the same hatcheries viz. TBD, BRP and KRP and between two different hatcheries mentioned above. These studies have shown that crosses between the individuals of the same hatchery did not show any clear differences in growth performance suggesting that no one strain would necessarily be recommended over another.

On the other hand, inter hatchery crosses between the strains at TBD and BRP and the reciprocal cross when stocked under polyculture along with rohu and mrigal at BRP and in two commercial farm ponds and under monoculture in a pond at the Fisheries Research Station (Hassarghatta) have shown an apparent enhancement of culture performance in the crossbred fish over the pure intrahatchery crosses.

There was considerable increase ($P<0.001$) in yield of the interhatchery crosses compared to intrahatchery crosses ranging from 120–245%, indicating positive heterosis for growth in the interhatchery stocks. Thus, to summarise the findings of Eknath and Doyle (1990) and Basavaraju *et al.* (1997), it is evident that the hatchery practices indicated inbreeding depression and negative selection for growth in domesticated stocks of Indian major carps.

## Status of Genetic Research on Indian Major Carps in Other Countries

The most actively involved countries, besides India, on genetic research of Indian major carps are Bangladesh, Vietnam and Thailand. Information is generally lacking regarding genetic work in Pakistan and Burma.

### *Bangladesh*

Research on genome manipulation is being carried out and successful induction of meiotic and mitotic gynogenesis was achieved at the Fisheries Research Institute Freshwater Station, Mymensingh. Clonal lines of *L. rohita* produced by meiotic gynogenesis in the F generation were reported to have been produced. Bangladesh has also initiated selective breeding work on *C. catla* for genetic improvement. For this, catla were collected from different river systems and genetic characterisation is being carried out by applying electrophoresis and morphological methods.

As indicated by the surveys conducted in Indian hatcheries (Eknath and 1990 and Basavaraju *et al.*, 1997), the investigations in Bangladesh have more or less shown the same trend. The existence of 400 carp seed producing hatcheries has been reported from Bangladesh. These hatcheries have been engaged in the production of seed since the 1970s. Some 31 billion spawn are reportedly produced annually. Problems mentioned with regard to hatchery produced seed in India have been also common in Bangladesh. Deterioration of hatchery bred stocks as indicated by retarded growth, morphological deformities and increased incidence of disease has been reported in recent years due to poor brood stock management. The data collected from several fish hatcheries located in Jessore, Comilla and Mymensingh revealed that most of the hatchery operators have no basic idea about brood stock management such as recruitment of new breeders at periodic intervals in the stock, optimum density of the brood stock and their proper feeding (Hussain and Mazid, 1997). These hatchery procedures lead to unconscious negative selection due to the use of undesirable size of brood fish and generation after generation of mating the related spawners from a small population.

Currently, Bangladesh is engaged in characterisation of different land races of catla and rohu by the assessment of meristic and morphological characters, genetic fingerprinting of riverine stocks of catla using molecular techniques and karyotyping to assess ploidy of rohu and mrigal. Developing breeding plans for stock improvement of catla and induction of gynogenesis and production of genetic clones in rohu are part of a genetic improvement programme for carps in Bangladesh.

### Thailand

Thailand has an elaborate genetic improvement programme, not only on carps but also for other species such as tilapias, catfishes, freshwater prawns, seabass, oyster, and aquarium fish. Among Indian major carps *L. rohita* has been currently taken up in Thailand for genetic work. Nineteen pairs in 1968 and 250 fish in 1969 were introduced into the country. The species is propagated through hatchery breeding and stocking in reservoirs and swamps where they have been already become established. It is also cultured in ponds. The production of rohu in 1994 was estimated at 1,783 tones ranking third in both weight and value among the carp species cultured in Thailand.

Genetic studies on rohu have been carried out with regard to the growth rate of the species raised in cages and to estimate the heritability as well as the genetic, phenotypic and environmental correlation for body weight, total length, body depth, head length and girth of fish at different ages.

Mrigal was introduced into Thailand during 1980 and 82. A consignment of 100 fish was first introduced from Bangladesh. Later in August 1982 eight pairs of brood fish weighing 700–800 g and 500 fingerlings were brought to Kalasin Inland Fisheries Station. Like rohu the propagation of mrigal is also carried out through hatchery breeding and the seed is stocked in reservoirs and swamps. The species has now become established and is available throughout the country. Thailand wants to set up a National Breeding Programme for hatchery produced carps and tilapias; among the Indian carps, *L. rohita* has been selected. The Central Institute of Freshwater Aquaculture (CIFA) is providing the improved rohu strain under the International Network for Genetics in Aquaculture (INGA) programme.

### Vietnam

Vietnam, like India, Thailand and Bangladesh, is one of the member countries of INGA but also the carp genetic improvement programme in Asia. The Indian carp species included in carp culture practices in Vietnam are the rohu and mrigal.

Vietnam has an elaborate selective breeding programme for mrigal. Mrigal production is mainly from north Vietnam, but this species is not considered as important in the south. Of the 200,000 tons production from freshwater, mrigal contribution was reported to be 45,000 tons in north Vietnam. In north Vietnam, sexual maturity in mrigal occurs at two years of age as in India. The breeding process is repeated three times in a season. It has been reported that the first and the second breeding yield better quality seed than the third. This is because young and small brooders normally mature and spawn earlier in the season and, like in other regions, early produced fry fetch higher price. This tendency to prefer to use early mating and small sized brood stock may lead to selection for small size and early maturity.

The base population of mrigal consisted of about 1000 fingerlings, imported from only one source, i.e. the Thangon state fish Farm in Laos in 1984 to the Thu Due Fish Farm of Research Institute for Aquaculture No. 2 (RIA 2) with the support of the Interim Mekong Committee. Another consignment of 10 pairs of breeders were imported from the same source in 1986. At the Fish Farm of the Research Institute for Aquaculture No. 1 (RIA-1) about 4–5 generations have been produced from only 6 males and 6 females which were transported from RIA-2 from the original stock. Because of the repeated production from a limited stock of 6/6 brood fish, the growth rate was observed to decline year by year. Deformities have been consistently observed in about 5–7 percent of the fish. The fish also have been reported to mature at lower age and lower body weight.

Vietnam is planning to take up individual (mass) selective breeding programme for mrigal. However, before the start of the programme, the base population has to be upgraded. Thus, Vietnam has planned to directly import wild populations of mrigal from three different river systems in India, consisting of about 2000 fry from each river. The two main goals of Vietnam for aquaculture of mrigal carp are a high growth rate and a high proportion of females, as males are 60–70 percent of the female body weight at market. This is true in most other carps as well A detailed plan of the selection programme for mrigal in Vietnam has been drawn out by Bentsen *et al.* (1996).

## Cryogenic Preservation of Carp Milt and Gene Banking

Protocols have been developed for sperm cryopreservation of major carps more or less simultaneously at the National Bureau of Fish Genetic Resource (NBFGR) and The Central Institute of Freshwater Aquaculture (CIFA) in 1988. The optimum requirements have been developed in respect to extender, activation media, dilution rate, activation period, sperm and egg ratio difference between species. Stocks of *L.rohita* milt cryopreserved in 1988 are being maintained along with other species in a mini gene bank at NBFGR. Viable hatchlings have been produced from five-year old rohu cryopreserved milt (Ponniah, 1997).

The ionic composition of seminal and variation fluid of rohu and mrigal have been studied at the Bureau with the objective of providing basic information that can be used to fashion new extenders and spermatozoa motility inducing media. Gupta and Rath (1993) preserved the milt of *C. catla, L. rohita* and *Cirrhinus mrigala* under cryogenic condition for three years. The quality of milt samples were analysed when preserved and were expressed in terms of spermatocrit value and sperm cell count, which in catla, rohu and mrigal ranged from 65–75, 75–85 and 65–75 respectively while sperm cell counts ranged between 2–2.5 X $10^7$, 3.0–3.25 × $10^7$ and 2.0–2.5 X $10^7$ respectively. The most effective diluent was found to be an electrolyte medium when Dimethyl Sulfoxide (DMSO) acted as cryoprotectant and the ratio of milt and diluent was 1:4. Quick Thawing procedure (38+2°C).

Gupta and Rath (1995) have also studied the efficacy of cryopreserved milt in fertilizing the eggs compared with normal milt.

The survival and growth rates from larvae to fry and to fingerling produced from eggs fertilized with cryopreserved and fresh milt were not found statistically different.

The fertilization efficiency of cryopreserved spermatozoa thawed at different temperatures of catla, rohu and mrigal over a period of four years was assessed at different time intervals after thawing, (Gupta *et al.*, 1995). Milt thawed at 39 + 1C and 30 + 1C had fertilization rates of 56.7% and 12.5% during 900 and 1500 seconds. The motility and fertility were reduced drastically at the next 60 seconds and 300 seconds.

## Cryopreservation of Embryos

Cryopreservation of embryos of *Cyprinus carpio* and *L. rohita* have been attempted (John *et al.*, 1993). It has been reported that *C. carpio* at tail bud stage was more tolerant to cryoprotectants than morula stage as revealed by the hatching rates of treated embryos. This may be due to some ontogenic differences in the tolerance to handling, membrane permeability or osmoregulation potential, rather than cryoprotectant susceptibility. Rohu tail-bud embryos treated in methanol and DMSO (Cryoprotectents) in different concentrations and in combination, showed high hatching rates in all the treatments but no pattern was evident. When embryos from all treatments were kept immersed in liquid nitrogen for about 3 hours and thawed, no survival was observed. This was felt due to defective thawing (John *et al.*, 1993).

Cryoprotectant toxicities have been also worked out in two embryonic stages of prawns viz. *Penaeus monodon* and *P. indicus*. Success has been achieved in obtaining viable larvae of *P. indicus* frozen to -40C and of *P. monodon* frozen to -30C. The frozen *P. monodon* larvae developed further without any abnormalities. Also, safe levels of cryoprotectants have been worked out for the embryos of *L. rohita* and common carp. Further research to develop, vitrification and cryopreservation of carp embryos is in progress (Ponniah, 1997).

## Gene Banking

The development of effective protocols for cryopreservation of sperm and embryos will go a long way in support of sustainability in aquaculture. It will also facilitate the establishment of a national facility for gene banking of fish gametes, embryos and DNA material.

The existence of such gene bank facilities will be also highly helpful for the obvious reasons that it reduces the maintenance problems and costs involved in raising brood fish in *in situ* condition, provides a wider choice of genetic material for selective breeding, and enables the process of dissemination of genetically improved varieties of fish with ease. Conservation of germplasm of endangered and wild populations would become easier.

As mentioned earlier the mini gene bank at NBFGR can be upgraded into a National Fish Germplasm Bank for wild and endangered species. Similar gene banking facilities can be also developed at CIFA for preservation of genetically improved carp and other domesticated culture species.

## Breeding Fish

At AC Tropical fish breeding you will find a lot of information and guidelines that can be useful when you attempt to breed your aquarium fish, or when you suddenly realise that your aquarium is filled with eggs or fry. The level of the articles varies from very basic introductory articles to more advanced accounts regarding the breeding of more difficult aquarium species. If you do not find what you are looking for on this page, we suggest that you take a look in the species database as well since a majority of the species profiles contain breeding information.

Some fish will readily spawn in aquariums while others are very difficult to coax into breeding in captivity but this is on of the things that make fish breeding so interesting. There are also many aquarium species that have never been bred in aquariums and where each specimen has to be caught from the wild.

Fish are either livebearers or egg layers. Livebearing fish give birth to free swimming fry and the offspring develops inside the female fish. Egg layers will instead release eggs that are fertilized outside the female fish.

Eventually the eggs will hatch and fry will emerge. If you are a beginner aquarist and want to try breeding fish, livebearers are usually a better choice than egg layers, since there are several livebearing species that will happily spawn in aquariums even without any extraordinary care or treatment. Molly, Guppy and Platy are three examples of commonly kept livebearers.

Livebearers give birth to fry that look like tiny copies of adult fish. It is however common that the fry is protected by a duller and more camouflaging colouration and that they lack extravagant features that are found in mature fish.

During the breeding period, the male will use his gonopodium to fertilize the eggs inside the female. The gonopodium is actually the anal fin of the male fish that has developed into a reproductive organ. After fertilization it will typically take four weeks before the fry is born. The fry is not born until it is developed enough to survive without any help from the parents.

Different egg laying species have developed different breeding strategies. Some are devoted parents that build advanced nests for their offspring or even carry them around inside the mouth, while others will simply release the eggs and let the current sweep them away as soon as they have been fertilized. Marine fish species will often release eggs that can float and stay hidden among plankton.

Egg laying species are usually divided into four basic groups: egg scatters, substrate spawners, bubble nest builders and mouth brooders. Fish species that live in rivers and streams are often egg scatters and they will typically let the current transport the eggs away and distribute them over a large area.

The egg can also fall down to the bottom where the fry develop. Since egg scatters hardly ever encounter their offspring in the wild, they are likely to eat them in the aquarium since the enclosed space is an unnatural living environment. Substrate spawners will instead attach the eggs to some type of substrate to make it impossible for currents to carry them away.

Some species will only use one specific type of substrate, and sometimes also vigorously clean it beforehand, while others are less selective. It is common for the male and/or the female to create a form of nest in the substrate. Bubblenest breeders will instead create a nest that is floating far away from the substrate. In most bubblenest building species it is the male that is responsible for building the nest and he will blow out tiny bubbles that he carefully glues together with saliva. Mouth brooders carry eggs and/or fry inside the mouth of a parent. Some species are maternal mouthbrooders while others are paternal mouthbrooders. Mouthbrooding will often be combined with substrate spawning.

## Breeding Angelfish

### *Sexing Angelfish and Obtaining a Suitable Pair*

It is usually very hard to distinguish a female Scalare form a male one, since the look very similar. Most Angelfish keepers are actually unsure of the sex of their fish until they notice which one is laying the eggs and which one that is fertilizing them.

If you watch your Angelfish closely right before the spawning commence, you will however notice a small difference in the size of the papilla between the sexes. The papilla is a pink coloured organ formed between the anal fin and the ventral fin.

Both sexes will display a papilla right before spawning, but the Angelfish with the largest papilla is usually the female one. Her papilla will also be blunter than the males, and when the female Angelfish becomes gravid she will display a large bulge in the region around her papilla. You can also notice a difference in behaviour; a male Angelfish is more territorial than a female. If one of your female Angelfish becomes gravid, all the male Angelfish in the aquarium will form a papilla. Those fishes that do not display a papilla in this situation are either females or unhealthy males.

Angelfish forms pair before breeding. They do however not pair up for life. The easiest way of obtaining a pair is to buy an already established pair from a breeder or a fish store.

This is however quite expensive and forming your own pair is much cheaper. A recommended way of forming a pair is to buy 4-8 Scalare and place them in the same aquarium, where they can get to know each other and form their own pairs. Choose fish with good external features, that looks strong and that display a healthy appetite.

Study the fins and gills extra carefully for signs of illness. It is advisable to buy from an established breeder if you want to produce high quality Angelfish.

### *Getting Angelfish into Spawning Condition*

Your Scalare will usually need to become at least 5 cm / 2 inches before they are mature enough to spawn for the first time. If kept on a suitable diet, this will happen when the Angelfish is between 8 and 12 months old. To induce spawning, you should keep your Angelfish in an aquarium that is large enough and not overcrowded.

The aquarium should ideally be at least 40 cm / 16 inches high, since the Angelfish is a tall fish. An aquarium that is to shallow will stunt their growth.

Angelfish can survive quite poor water qualities and unsuitable water temperatures, but it can make them stop eating and will inhibit breeding. Frequent water changes should therefore be performed and the water temperature kept between 80 and 85 degrees F (27 and 29 degrees C). Suitable food is also necessary to get your Angelfish into spawning condition. The diet should be varied. Prepared foods especially made for Angelfish is a good base, but should be supplemented with meaty foods. Mosquito larvae, daphnia, beef heart, brine shrimp and larger brine shrimp cut up into suitable size are food types known to induce spawning in Angelfish.

Even though your Angelfish must be given plenty of food, you must also avoid overfeeding them. Check the aquarium 5 minutes after the feeding and remove any food that has not yet been consumed. If you frequently find left over food in the aquarium after 5 minutes, you are feeding your Angelfish to much.

### Spawning Signs in Angelfish

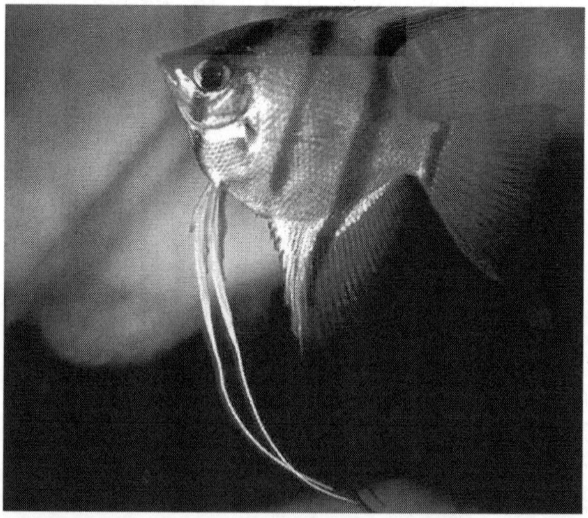

***Figure:*** *Choose good parents when breeding angelfish.*

If an Angelfish suddenly displays a bulging belly and a more aggressive behaviour, it is probably a gravid female Angelfish. Two Angelfish grooming each other is also a typical breeding behaviour. Angelfish of both sexes will also flash fins, face off, lock their mouths and twist around. When a couple has been formed, they will choose a spawning site and start cleaning it together.

### Angelfish Egg Laying and Fertilization

The Angelfish eggs will be deposited on the spawning slate and placed in very neat and evenly spaced lines. If she doesn't like the spawning slate, she will deposit the eggs somewhere else in the aquarium. The male Angelfish will fertilize the eggs by following close behind the female and touching all the eggs with his papilla. If you keep only female Angelfish, one of the females might actually resume a male behaviour and follow the egg-laying female and touch the eggs. Those eggs will naturally remain unfertilized and become white within a few days.

### Angelfish Egg and Fry Care

Several methods are used by Angelfish breeders to protect eggs and fry from hungry adult fish. Some remove all fish except the breeding pair from the aquarium, while other breeders choose to remove the breeding pair instead and place them in a special breeding aquarium. Some remove less dominant males until only the most dominant male is left with the females in the aquarium. This method also involves removing all non-gravid females from the aquarium.

All adult Angelfish pose a treat to the eggs, and many breeders will therefore place the eggs in their own tank as soon as the eggs have been fertilized. The easiest way is to put a special spawning slate in the aquarium before the spawning takes place. The parent fish will deposit the egg on the spawning slate, and the spawning slate can be easily moved afterwards. A clay pot, poly pipe or a piece of smooth surfaced tile will all work as spawning slates. The eggs will survive a brief moment out of the water when you lift them from one aquarium to the other, but try to make the move as fast as possible.

If you do not wish to keep all your fry, you can let the eggs stay in the aquarium with the adult fish. If you provide your fry with suitable hiding places, such as heavily planted areas, at least a few of them will reach adulthood. Be aware of strong filtration, since small fry can be sucked into filter systems.

If your Angelfish pair has never spawned before, it is usually a good idea to let them spawn once or twice in their normal aquarium "on their own" without your meddling. If you move them to a breeding aquarium during their first spawning they might be confused and stop the spawning process. It is normal for the parents to eat the eggs from their first spawning, and if this happens the female will usually deposit eggs again after just a few weeks.

### Angelfish Development

***Figure** : Angelfish breeding have result in many types of angelfish*

Newly deposited Scalare eggs are transparent or semi-transparent, with a pale yellow colouration. If the eggs are not fertilized, they will start turning white after approximately 24 hours. If they have not been eaten by other inhabitants of the aquarium, unfertilized Scalare eggs will be attacked by fungus and start looking very fuzzy after roughly 48 hours. 48 hours after being deposited, fertilized eggs on the other hand will begin showing wiggling fry tails.

On day three, tadpole shapes with large yolk sacks will be visible. The tadpoles will still be attached to the breeding slate. If the parent Angelfish are still in the same aquarium as the egg, the parents will now start to move the small fry around. On the fourth day you will be able to see tiny eyes on the fry and on the fifth day the yolks will start shrinking. Very soon the first fry will be free swimming. During day seven the fry will usually start getting hungry since their energy from the yolk have been consumed.

### Suitable Tank for Angelfish Fry

If you decide to raise the Angelfish eggs and fry in a special aquarium, a 5-10 gallon (19-38 litres) tank will be suitable, but even a tank no larger than 2 gallons will usually work. You do not need to cover the egg and fry aquarium. If you allow the parent Scalare to accompany their offspring in the aquarium, the parents will keep the eggs clean and protect them from mould.

If the eggs are kept without the parents, you must use fungicide to prevent fungus attacks. Make sure that the water temperature in the raising tank is the same as in the parent's old aquarium. An air stone in the raising aquarium will make it easier to maintain high oxygen levels. Keep the water quality good by frequently removing any dead fry and left over food. Regular water changes are also a must. Typically, a lot of the Scalare fry will die. You might have to remove dead fry from the aquarium three times a day or more. During the first three weeks, water changes can be necessary as often as once a day.

### Feeding Angelfish Fry

You can start feeding your Angelfish fry around 7 days after the eggs were deposited. Angelfish fry can survive 12 hours without food, but should ideally be fed four or five times a day. Newly hatched Brine Shrimp (Artemia nauplii) are a good first food for Angelfish fry. If the fry is too small to consume the Brine Shrimp, you can buy the fry special fry food from a pet shop or fish store. The survival rate for Angelfish fry is usually very much lover when the fry is fed dead food instead of live food (such as live Brine Shrimp). When the fry are a few weeks old it is time to make them grow accustomed to flake foods. Turn the flake food into a powder or press it through a fine screen before feeding it to the fry.

### Angelfish Guarding and Caring Behaviours

As mentioned above, it is quite common for Angelfish to eat their eggs or fry, especially during the first few spawnings. This is why most Angelfish breeders place the eggs in a special aquarium without any adult fish. It is however very fascinating to watch an Angelfish pair taking care of their own offspring. You can for instance remove some of the eggs, and let some stay with the parents. If you leave the eggs to hatch with the parents, the parents will spawn again after 2 or 3 weeks and the already hatched fry will use the eggs as food. When the fry is 5 or 6 weeks old you will see them eat from the skin and fins of their parents. This behaviour can make the parents very weak and even cause them to die.

Some Scalare individuals are more aggressive than others, and will not be able to raise their own offspring. They will always eat their offspring or at least chaise them away. Such individuals can usually be singled out as early as during the courtship period, since they are extremely aggressive and sometimes even injure a fish they wish to breed with. For instance, it is not that rare for Angelfish to suffer from severely injured mouths after the courtship period.

## Breeding Bristlenosed Catfish

Bristlenosed catfish is not very hard to breed in aquariums. Keep one male Bristlenosed catfish with several females, and make sure they are all in top condition. Provide them with good water quality and a varied and nutritious diet. You will most likely never notice the actual spawning behaviour, just the result of it in the form of tiny Bristlenosed catfish babies that suddenly appear in the aquarium.

From what we know about Bristlenosed catfish breeding habits, the spawning seems to be initiated by the male catfish. He will pick a suitable spawning site, usually a cave or a crevice, and subsequently start chasing the females as a part of the courting behaviour. The females will repeatedly inspect the spawning site and this part of the spawning can go on for several days. Eventually the female catfish will deposit her eggs at in a clump at the spawning site where they are fertilized by the male. Newly deposited eggs from a Bristlenosed catfish display a bright orange colour. The female will leave the spawning site immediately and do not stay around to guard or care for the eggs.

The male Bristlenosed catfish stays at the spawning site to protect the eggs from predators. The will place his tail near the eggs and let his head protrude out of the cave or crevice. He is probably placing his tail close to the eggs in order to care for them by fanning fresh water over them, but this has still not be determined. The male Bristlenosed catfish is a very capable parent and you should never remove him from the eggs or try to raise catfish eggs in a breeding tank without the male present.

Fry will emerge from the eggs after 4-5 days. The fry will still eat from their yolk-sac. They will start to spread over the aquarium, and in a community aquarium a lot of them will soon be consumed by the adult fish. If you want to ensure a high fry survival rate it is recommended to breed Bristlenosed catfish a separated tank. You can even remove the females from the aquarium, and leave the fry with the male. When the fry has been free swimming for a week you can remove the male as well. High oxygen levels are very important for Bristlenosed fry and good aeration in the aquarium will typically lead to a higher survival rate.

When the yolk-sac is consumed you can start feeding the Bristlenosed fry vegetable food, such as very finely cut lettuce and cucumber. Dip the vegetables in boiling water before you place them in the aquarium. A good algae growth in the aquarium is beneficial for the fry since this is their natural food source. If kept on a suitable diet the fry will be around 1 ¼ inches long after two months.

## Breeding Discus Fish

Discus is not a very easy fish to keep, and breeding them in aquariums is even harder. Only those with previous experience from keeping easier species should try caring for Discus, since many Discuses have died in the hands of inexperienced aquarists. Discus fish are wonderful to keep – they are beautiful as well as charming – but they do require quite a lot of pampering. The key to successfully breeding Discus lies in always providing them with supreme water conditions, keeping them in an aquarium that is large enough for them, giving them high quality food etcetera. A well kept Discus pair will breed regularly in the aquarium, and there is really no special "fix" that will induce breeding in a couple that do not feel well. Most of the guidelines regarding Discus breeding will therefore be the same as for keeping Discus.

First and foremost – a Discus couple will not breed in an aquarium that is too small. A depth of at least 15 inches is required for these tall fishes. A 36 inches x 18 inches x 18 inches aquarium is therefore usually the smallest suitable standard aquarium for a pair. If you have a 48 inches x 18 inches x 18 inches aquarium you can even keep a group of 4-6 smaller Discus until they have formed their own pairs.

The water temperature should be kept at 82 degrees Fahrenheit or higher, and the water must be very soft. Keep the pH as close to 6.5 as possible. The soft water means that you will have to monitor the pH value vigilantly. Soft water is a very bad buffer and the pH value can therefore drop very quickly and your Discus will not do well in acidic water. Keep the levels of soluble waste as low as possible and never allow the nitrate level to rise above 20 ppm. Since your

Discus will require a diet rich in protein, they will consequently produce a lot of waste products that will push up the nitrate levels in the aquarium. The best way to combat the nitrate is by performing frequent water changes. Changing 30-50 percent of the water each week is a good rule of thumb.

As mentioned above, your Discus will require plenty of protein in their diet. Professional Discus breeders usually give their fish beef heart. The beef heart diet should ideally be supplemented with other food types, such as spinach and bloodworms. Some high quality flake food will provide your Discus with extra nutrients, and additional vitamins can also be a good idea.

During the initial stage of the breeding process your Discus couple will start to defend a small region of the aquarium. Both the male and the female Discus will become highly territorial and guard this region together. They will continue by picking a spawning site and carefully clean it. Some Discus pairs clean a number of sites before they finally decide to place the eggs on one of them. The female Discus will swim over the spawning site and deposit her eggs in rows, and they male will swim behind her and promptly fertilize them. I wish to once again stress the importance of keeping your Discus in soft water. Hard water will make the protecting membranes that surround the eggs so hard that it is impossible for the sperm to get through. Hard water can therefore cause the eggs to stay unfertilized and no fry will develop. Hard water can also make it impossible for a fry to hatch.

Discus pairs are attentive parents and will guard and care for the eggs together. They will fan fresh water over the eggs to ensure high levels of oxygen and they will also remove any unfertilized eggs to prevent bacteria growth and mould. Discus eggs will hatch within 48 hours if the water temperature is right. The fry will stay at the spawning site with the parents for an additional 12 hours. After that, the fry will be moved around the aquarium by their parents, and the parents will use different places to hide the fry in. If you keep up the water temperature your will have free swimming fry in the aquarium after approximately one week. Free swimming fry will need no extra food, because they will eat from the scales of their parents. The adult Discus will secrete baby-food from between their scales. This food will provide the fry with everything they need for about one week. After that, they will be large enough to eat more normal food and you can start feeding them newly hatched brine shrimp. Powdered flake food is another option. Just like their parents, Discus fry will appreciate some beef heart, but since the fry are so small you will need to liquidise the beef heart for them. Do not separate the fry from the parents until you know for sure that they are no longer feeding from the scales. If you provide your Discus fry with ideal conditions and suitable food, they will grow very rapidly and be around 2 inches in diametre at an age of 12 weeks. When they have reached this size, they are large enough to be sold to pet shops or other aquarists.

The parents are extremely important when it comes to keeping eggs and fry alive and it will be practically impossible for you to raise Discus without the parents. Do not be discouraged if you Discus pair fail to raise their first batch successfully. The mere fact that they are spawning in the aquarium is a good sign, a one failed spawning will usually be followed by a more successful one eventually. It can take a Discus couple numerous spawnings before they finally get it right.

## Breeding Goldfish

Goldfish cover a wide range of species, but they are all members of the generic family Carassius auratus, and have been bred by the Chinese, Koreans and Japanese. One of the difficulties for the amateur to breed this fish is that they can be difficult to sex. The difference in their gender is only apparent when they are in a spawning condition. The greatest change occurs in the males, they develop white pimples known as "breeding tubercles", on their fins and sometimes their head. Check along the gill operculum, and along the front rays of their pectoral fin.

Goldfish are capable of breeding when they are a year old, but they are in their prime during their third year. Under normal circumstances, it is impossible to tell their sex until they have grown to 8cm. Therefore it is impossible to buy a small fish and be certain of its gender. Many breeders advocate that the sexes should be kept apart before the breeding commences, and this is impossible before they are sexed. Therefore breeding of goldfish can be a long term project, as they first have to be sexed. This is best done in the Summer, and the fish can be over wintered for breeding in the Spring.

The process of breeding is induced by a change in the temperature of the water. Goldfish will breed at a temperature of 10-26°C, but the optimum temperature to ensure success is 20°C. To induce the fish to spawn you can either wait for more clement weather, or you can artificially and slowly raise the temperature of the water. To ensure a greater chance of fertility more males are kept then females, usually the ratio is two to one. Bear in mind here that each fish needs fifteen gallons of water, and they do not like to be overcrowded.

It is easy to spot when they are ready to breed, as the male has a ritual of a "spawning chase", whereby he follows the female around and pushes against her abdomen, for several hours. This should occur if all the conditions are correct with a few days of the fish being brought together. The female, when she is ready will lay her eggs in batches, the first batch being the most fertile. She can lay from five hundred to a thousand eggs, so they will not be released all at once. The male will immediately fertilize these eggs by spraying milt on to them.

Once the egg laying is over the eggs should be removed from the breeding tank, and rinsed in aged water, which is at the same temperature as the water they have been released from. They should then be placed in a separate twenty gallon tank. However the depth of the water should not exceed six inches, as the pressure of the water inhibits their ability to rise to the surface.

The eggs will hatch at the temperature of 21°C, so if the water is cooler raise it gently to that temperature. At this stage the clearer looking of the eggs are the most fertile, and therefore have the greater chance of hatching. The darker eggs are infertile, and have a high chance of developing a fungal infection; they should be removed from the tank. A few drops of methylene blue can reduce the chance of fungal activity.

***Figure:*** *Goldfish breeding have resulted in many types of goldfish over the years*

Once the fry have been hatched they have a full yolk sac, and they should not beefed during the first forty eight hours. During this time it is normal that they will sink to the bottom of the tank, and they should be left undisturbed. During this period they are absorbing the yolk sac and developing an air bladder. Once the air bladder has developed, they will rise from the bottom naturally.

After forty eight hours they are hungry and ready to be fed. They may be fed from a commercial product or from a paste of oatmeal, and also hard boiled eggs. They need only sufficient food, which they can eat during a twenty minute period, after this time any uneaten food should be carefully siphoned off.

## Breeding Mbuna Cichlids

Mbuna cichlids are native to Lake Malawi in Africa. The word Mbuna means rock dweller in one of the local languages spoken around the lake, and it is truly a suitable name for these rock living cichlids. Mbuna cichlids are popular in aquariums due to their beautiful look – most species display very striking colours – and their fascinating behaviour.

They are very aggressive and should not be kept with more timid fish. Mbuna cichlids will claim territories and defend them violently. They should ideally be kept in a well stocked aquarium, since this will cause the aggression to be spread out over many individuals. One inch of fish for every 8 square inches of water surface is a good rule of thumb when it comes to Mbuna cichlids.

It is not hard to get Mbuna cichlids into spawning condition. Just provide them with the same care and conditions as you would provide non-breeding Mbuna cichlids with. This article will provide general guidelines that are true for most Mbuna cichlid species, but it is always advisable to research your particular species as well before you attempt to breed them.

Also remember that Mbuna cichlids from different species can and will cross breed. If you keep several species together and they produce offspring you should not sell the fry since the mixed DNA will ruin the bloodlines of the true species when people use the crossings for breeding purposes.

Keeping one male with at least 2-3 females is a good ratio. As mentioned above, the Mbuna cichlids should be kept in a well stocked aquarium. This means that you might have to add extra aeration and filtration, and perform water changes quite often. Your Mbuna cichlids will not breed if you allow the water quality to trop. Lake Malawi is a large lake with a stable environment, and Mbuna cichlids to not appreciate rapid changes. Keep the pH in the 7.5-8.5 range and the KH as well as the GH at 12.

The levels of ammonia and nitrite should be zero, and the nitrate should never rise above 15 ppm. The normal temperature in Lake Malawi is 75-80 degrees Fahrenheit.

If you want to breed you Mbuna cichlids, it is important not to over feed them. They will eat virtually everything you give them and can easily become obese which is bad for their health. Very large Mbuna cichlids also means that you will be able to keep fewer individuals in your aquarium, and the amount of aggressive behaviour will therefore increase. Violence will make the fish stressed and they are less likely to breed under such conditions.

In the same time, it is important not to overcrowd the aquarium, since this will hamper the breeding as well. The natural diet for Mbuna cichlids are algae and insects. If you feed you Mbuna cichlids prepared food, you should always include some vegetable/algae matter such as Spirulina.

Since Mbuna cichlids are rock dwellers, it comes as no surprise that they like to spawn among rocks, caves and crevices. Mbuna cichlids should never be kept in an aquarium without plenty of hiding spaces created by rocky decorations, and if you want them to breed this is even more important.

During the initial stage of the breeding period the male will claim a small territory which he thinks is a suitable spawning site. He will then start to show him self in front of the females. He will 'dance' for them by vibrating his whole body and erecting his fins. His colours will also become much more vivid than normally. If a female is in spawning condition, she will eventually accompany the male to the spawning site.

The couple will swim close to each other in a circling fashion, and the female will start depositing eggs at the spawning site. When a few eggs have been deposited, she will quickly pick them up with her mouth to guard them. Mbuna cichlids are female mouth brooders and will keep their offspring hidden inside the mouth of the mother.

The male is equipped with some egg-shaped spots on his anal fin, and when the female Mbuna tries to pick them up as well she will instead receive a mouthful of sperm that fertilizes the eggs inside her mouth. She will then continue to deposit eggs at the spawning site, and the procedure is repeated over and over again until all the eggs are fertilized and resting inside her mouth.

When all eggs has been fertilized, the male Mbuna will chase away the female from his territory. It is unadvisable to remove the female from the aquarium at this stage, since the stress from being netted and moved can cause her to spit out the eggs. If you have filled the aquarium with a lot of small hiding places for the fry to seek refuge in, you can let her stay in the aquarium and let the fry grow up together with the adult fish. Some fry will be eaten, but at lest a few will survive. Their instinct will make them seek shelter among rocks and stones just like they would do in the wild. If you want to ensure a higher survival rate, you can move the female to her own aquarium approximately 7-10 days after spawning. She is less likely to spit her eggs when she has kept them in her mouth for such a long time.

The fry is usually spit out after 4 weeks of brooding, but you should not move the female later than 10 days after the spawning since a late move can hurt the fry.

When the Mbuna fry are released they are already comparatively large and you can feed them Cyclops, brine shrimp or powdered flake food. One batch of Mbuna fry will typically include 25-35 individuals. Let the female stay with her fry for a while and give her plenty of nutritious food. She needs some time to regain her strength before you move her.

## Breeding Oscar Fish

Oscar fish can be bred in aquariums as long as you provide them with favourable conditions. To begin with, the aquarium must be large enough. Oscars are large cichlids and their size can exceed one foot. To get a breeding pair, you must house 6 young Oscars together and let them form their own pairs. Make sure you choose healthy Oscars from different broods. It can be hard to sex young Oscars, but you will hopefully get at least one pair from 6 individuals. If you don't want to wait and let Oscars grow up together, you can instead buy a breeding pair but this is usually much more expensive. The stress and disturbance experienced by the fish when you move them from their old aquarium can also cause them to brake up and stop being a breeding pair.

Once a pair has been formed you should remove the other fish from the aquarium, since breeding Oscars are highly territorial and can hurt intruding fish. If the spawning does not commence immediately there is not easy fix that can kick start it. What you can do is provide the pair with suitable conditions and wait for the spawning to start.

The courtship behaviour can be very violent – sometimes too violent and resulting in the death of one of the Oscars before any offspring has been produced. Oscars will do well in soft acidic water as well as in harder alkaline water, as long as you avoid any extremes. Keeping the conditions similar to their native home in Central America is always a good idea. Frequent water changes are important to ensure suitable water chemistry and low levels of soluble waste. Do not change more than 25 percent of the water at once. Changing 20-25 percent of the water once or twice a week is a good rule of thumb. Efficient filtration is also required since Oscars are renowned for being messy eaters and producing plenty of waste. If you have a 50 gallon aquarium you should ideally use filtration that is recommended for a 100 gallon aquarium.

Feed the pair bait fish or frozen fish. Prawns, crickets, earth worms and other meaty foods are also suitable. Some Oscar keepers feed their fish raw meat from cows and pigs, but this is highly unadvisable since such meat include too much animal fat and can harm the health of your Oscars.

Remember that wild Oscars feed mainly on small fish and eating cows, pigs, sheep etc are therefore very unnatural for them. Pellets should also be avoided since a majority of the pellet will slip through the gills of the Oscar and never end up inside the stomach. Your fish will still be hungry and the water will be polluted with uneaten pellet material.

An Oscar pair will typically dig a pit in the substrate or choose a hard surface, such as a flat stone, as their spawning site. Both the male and the female Oscar will guard and care for eggs and fry.

The will be extremely protective and aggressive during this period and should be kept alone. Just like with many other cichlid species, it is common for an Oscar pair to fail their first attempts to raise offspring.

Newly hatched brine shrimp is a suitable first food for Oscar fry. You can feed the powdered flake food, but they will grow much faster on brine shrimp. When kept on a brine shrimp diet the fry will be 1.5-2.0 inches long after three months. Keep in mind that one Oscar pair is capable of producing tremendous amounts of offspring. You might have to cull the batch if you can't provide all the young Oscars with suitable housing.

## Breeding Tiger Barbs

Tiger Barbs are a good choice for beginner aquarists that wish to breed fish, since Tiger Barbs frequently spawn in aquariums without much pampering. They are hardy and will accept a wide range of temperatures. If you choose a high quality flake food they will do well on a prepared food diet, but it is always recommended to sometimes supplement the flakes with more meaty foods such as Daphnia, bloodworms or similar.

The Tiger Barb is an active fish that can reach a size of 1.5", so make sure you have enough space to keep the produced fish in before attempting to breed them. Since Tiger Barbs do not raise their offspring, one Tiger Barb batch is usually very large to make sure that at least a few of the fry survive. This means that you might have to cull the batch if you do not have a very large aquarium for the fry to grow in. If you refrain from culling and instead let them grow up in a crowded aquarium, the whole batch will suffer.

If you want to breed Tiger Barbs, you must start with separating the males from the females. It is true that healthy Tiger Barbs sometimes spawn when kept together in community aquariums, but this it is not something that happens very often and if you want to induce spawning you should separate the sexes from each other. The smallest Tiger Barbs with the plumpest bodies are the females.

When you have separated the sexes you should start feeding them a spawning-inducing diet. Tiger Barbs are happy eaters and will usually consume anything you give them, but breeding Tiger Barbs require a diet rich in protein to do well. Freeze dried Tubifex is known to induce spawning in Tiger Barbs, but other protein rich food types can also work well. Live food is always a good idea.

The temperature in the aquariums where you keep your Tiger Barbs should stay around 80 degrees Fahrenheit. After 3-4 days of separation you should let the Tiger Barbs see each other again. The aquarium where you want them to breed must have slightly soft water, and the pH should be a bit acidic.

Just like when they were separated, the water temperature should be kept close to 80 degrees Fahrenheit. A mature sponge filter will help you keep the water conditions optimal. You must also provide your Tiger Barbs with suitable spawning sites, such as two bunches of Cabomba plants or a pair of spawning mops.

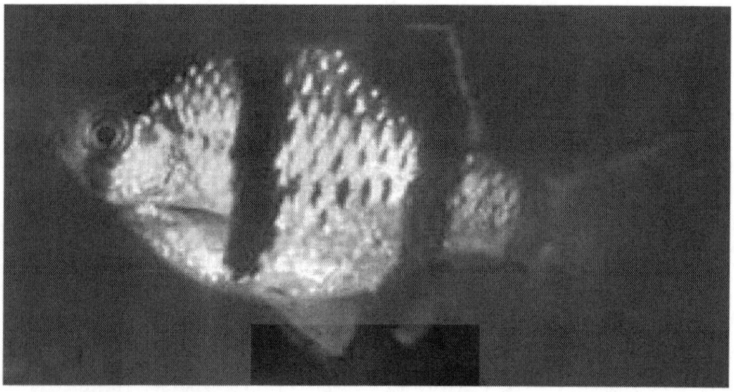

***Figure:** A Tiger Barb in breeding condition.*

Spawning will usually begin very soon when the Tiger Barbs have been brought together again. During the courting period, the male will chase the females. When a female Tiger Barb starts releasing her eggs at the spawning site the male will follow her and leave a trail of sperm in the water.

When the spawning is over you should remove the adult fish from the rearing aquarium, since Tiger Barbs are known to eat their own offspring. If you have covered the aquarium floor with marbles the eggs will be safer since the eggs can fall down between the marbles where the adult fish can't reach them. If you keep the water temperature in the aquarium at 80 degrees Fahrenheit you can expect free swimming fry after approximately five days. The fry will usually cling to the sides of the aquarium.

Newly hatched brine shrimp is a suitable first food for Tiger Barb fry. Powdered flake food will also work, but the growth rate will be slower and powdered flake food is not recommended for larger Tiger

Barb fry. If the newly hatched Tiger Barbs are very small, you can feed them liquid fry food until they are big enough to eat solid food. Tiger Barbs are commonly sold to pet stores when they are around three months old.

## Breeding/Rearing Aquarium Tanks

Some fish species will successfully breed in a community aquarium or species tank, but others will require a separate breeding aquarium. In some cases, the fish will spawn their normal aquarium but the fry must be caught and placed in a separate rearing aquarium to prevent the fry from being eaten by the adult fish. Fry left with adult fish will usually be consumed, but you can increase the survival rate by providing the fry with plenty of hiding spaces among marbles, densely planted bushy plants etcetera.

If you decide to setup a breeding/rearing tank, a 2-6 gallon container will be large enough for most aquarium species. Very large fish might require 10-22 gallons. The easiest way to fill your breeding tank with suitable water is to fill half the tank using water from your large aquarium where the adult fish live. Use clean water to fill up the rest of the breeding tank.

This water should be treated in the usual way for water changes. Another method is to place a foam filter in your large aquarium and let it run for a few weeks before you set up the breeding tank. This way, the foam filter will pick up a lot of necessary bacteria and the biological process in your breeding tank will get a kick-start when you insert the foam filter. A third method is to buy commercially available products such as Bio-Starter or Safe Water from your fish store and use them in the breeding tank. Some breeders combine the methods described above.

It is not recommended to place a lot of decorations in the breeding tank, unless you know for a fact that your particular species require some special type of decoration. A thin layer of gravel is easier to keep clean than a thick one, and cleanliness is very important in a rearing aquarium. A foam filter is the best type of filtration in a breeding tank since the filtration is gentle but efficient. The air pump must be of correct size for your breeding tank, or an adjustable model.

Even with a good filter, regular water changes are absolutely imperative in a rearing tank. High levels of soluble waste can harm and even kill eggs and fry. If there are no fry or adult fish in the aquarium, water changes can be performed much less frequent since eggs hardly produce any waste products. When the fry is very small, siphoning out waste products and left over food using a flexible tube is usually the safest cleaning method. Fish fry are renowned for their curiosity and accidents are common. Let the other end of the flexible tube rest in a clean bucket (do not clean with soap, detergent or similar!) when you siphon. This way, you can rescue the fry that becomes siphoned up and put them back in the rearing aquarium.

Plastic breeding traps are popular to separate fry from adult fish in community aquariums and species aquariums, but are not suitable homes for the fry for any length of time. The circulation of water is usually far from optimal and the fry should therefore be moved to a rearing aquarium as soon as possible. Another problem with plastic breeding traps is that many models float directly under the aquarium lights, which makes the water inside unsuitable warm for the fry. A breeding net that can be attached to the side of the aquarium is a better solution if you can't setup a separate rearing tank for the fry.

# 6

# Conditioning the Parent Fish

Before placing the parent fish together for spawning, they should be conditioned. Conditioning is feeding the fish a variety of healthy foods to get them in top condition for spawning. Many species can be conditioned using a well-balanced flake food, though others should be conditioned on live foods such as brine shrimp, insect larvae, and flying insects. The parent fish can be while conditioning. This way, when the fish are reintroduced, they are eager to spawn.

## Spawning Tank

Though some species readily spawn in the aquarium, the eggs or fry often do not survive because of predatory parents or other fish. Often the fry die because of unfavourable, unclean water conditions. Many species that practice brood care will harm other tank mates in attempting to guard the eggs. Because of all these problems; most aquariasts who breed fish use a separate spawning tank. The spawning tank should be like the hospital tank with protected heater so the the fish are not burned; a slow-moving filter (sponge filter), so the eggs or fry are not sucked up; and good aeration. Depending on the spawning method, the spawning tank can be set up in a number of different ways.

***Egg-scatterers:*** Because egg scatterers often eat their own eggs, the spawning tank has to be set-up so the eggs fall out of the reach of hungry parents. A 5-10 gallon tank is sufficient for spawning for most eggs scatterers.

For egg scatterers like barbs and danios, which lay non-adhesive eggs, the spawning tank can be furnished with a substrate consisting of two layers of marbles or a nylon netting just above the tank floor. As the eggs are laid, they fall through the marbles or the netting out of the reach of the parents. After spawning is over, the eggs or the parents can be removed.

For egg scatterers that lay adhesive eggs like tetras, the spawning tank should be furnished with a substrate. The tank should be planted with fine-leafed plants. The eggs are laid amongst plants, and adhere to the fine-leaves. The parents should be removed after spawning. Rainbowfish also lay adhesive eggs, however, most species spawn continuously over a period of several weeks. Thus, the eggs or the plants that the eggs are attached to, should be removed daily and placed in the rearing tank. Larger rainbowfish require a spawning tank of 20-30 gallons.

***Egg-depositors:*** Depending on the type of egg depositor, the tank should be furnished differently. For those egg-depositors that care for their young, the parents can remain in the tank after spawning. Substrate spawners, depending on the species, should be given a tank furnished glass panes, broad-leafed plants, or flat stones for spawning sites. Some species such as Discus and Angelfish prefer vertical surfaces. For cavity spawners, flower pots turned on their side, coconut shells, and rocky caves are suitable spawning sites. The tank should be furnished with either live or plastic plants to give the fish a sense of security.

Egg-depositors that do not care for their young, should be given a tank furnished with fine and broad-leafed plants, Java Moss, or artificial spawning mops. After spawning the parents or plants with the eggs should be removed. If the plants containing eggs are removed, new plants should be placed in the tank for future spawnings. Killifish eggs often develop best when they are kept in a shallow dish.

***Egg-burriers:*** A peat-moss substrate is one of the best substrates for egg-burying species. The peat moss can be removed after spawning and placed in a plastic bag to be stored for weeks to months (depending on the species). A new peat moss substrate can be placed in the tank for further spawnings. In order to initiate hatching, the stored peat can be immersed in soft water.

***Mouth-brooders:*** Ovophile mouth-brooders can be bred in the main aquarium because the eggs are protected in the mouth cavity. However, it is better to separate mouth-brooders with eggs because of their potentially aggressive behaviour. There are no special breeding tank requirements other than the usual tank set-up for the species.

Larvophile mouth-brooders should be placed in a breeding tank because the eggs are not protected in the mouth, but laid on a surface where the are open to predators.

***Nest-builders:*** Nest-builders should be provided with material with which to build their nests. For bubble-nest builders, fine leafed and floating plants should be provided, and the tank should have no water current to disturb the nest. Species that build nests in the substrate should be given fine gravel or sand.

***Livebearers:*** Small livebearers can be bred in breeding traps where the newborns fall out of the reach of the mother.

However, a more preferable set-up is a separate, heavily planted tank. As the female releases the young, she can be fed so that she is not to eat the fry. As soon as all the young are born, remove the mother.

## Stimulating Spawning

One of the best ways to induce fish to spawn, especially difficult-to-spawn species, is to simulate natural conditions. Among factors that encourage fish to spawn are the environment, the food, and the rainy season.

***Water Conditions:*** The right water conditions are among the most basic requirements in spawning fish. Thus the water conditions should be similar to those in the natural environment of the species. By following the suggestions under "breeding" or "water" in the species descriptions, approximate natural water conditions can be found. Another important environmental conditions is the the right tank set-up including hiding places, spawning sites, lighting, water current, and social conditions (schools).

***Food:*** The right foods are important to encouraging spawning. Without proper foods, natural conditions cannot be entirely recreated. Some of the live foods that often can make a difference in spawning success are mosquito larvae and fruit flies.

***Stimulating the Rainy Season:*** Many fish species spawn during the rainy season in nature. By simulating the rainy season in aquaria, difficult-to-spawn species can be induced to spawn. Rains affect the water chemistry, the water height, and the water temperature.

In order to simulate rainy conditions, the tank level should be reduced to half its normal height. Each day add 5% of the tank volume. The water added should be very soft and slightly cooler than the tank temperature. To simulate the rain, use a drip system or spray bar for several hours a day and simulate storms by flashing light in a dark room. Thunder can be created by playing a recording of a storm, a recording of music with bass, or a recording of rattling thin metal cake pans.

The process can be repeated until the fish show signs of spawning preparation. In addition to the "storms," the fish should be fed heavily with mosquito larvae, shrimp, and flying insects. This process will stimulate some hard-to-spawn species, but not all. Many species will respond to just one altered water condition, such as a series of water changes with soft water replacement, or a decrease in the water temperature.

***Target Fish:*** "Target fish" can be used to help strengthen the bond between a fish pair. Target fish can be another of the same species or a similar species that is placed in the tank with the breeding pair. This third fish will serve as an object of the aggression of the pair. The pair will work together to chase off the target fish and not fight between themselves. Only use the target fish method in a large tank with plenty of hiding places, so that the target fish is not harmed.

## The Rearing Tank

A rearing tank is not required with species that take care of their young, although they are still recommended. For species that do not take care of their young, the rearing tank can be the same tank as the spawning tank as long as the parents are removed.

The rearing tank should have a protected heater, a sponge filter or a filter with a sponge on the intake, and plastic or live plants. In addition, three of the tank sides should be covered with black paper, because a light can encourage fungal infections and cause discomfort for the fry. The water in the rearing tank should be similar to the water used for spawning.

## Raising the Fry

The fry develop in a number of minutes (livebearers) to a number of weeks. When the eggs hatch, the larvae that emerge look nothing like the parent fish. Instead, the larvae have a large, yellow yolk sac and are barely able to move, let alone swim. The larvae will feed off the egg sac until all the yolk is gone. Once the yolk sac is gone, the hungry fry will begin to look for food. The fry of small fish can be first fed infusoria, "green water," or egg yolk. Later these fry can be fed larger foods like whiteworms, Daphnia, Artemia nauplii, and ground flakes. These foods are good as a first food for slightly larger fry such as those of cichlids. Once the fish grow larger, larger foods like brine shrimp, larger Daphnia, flakes, insect larvae, and chopped Tubifex worms are accepted.

25-50% of the water in the rearing tank should be changed daily. Be sure that the "new" water added has characteristics like the water taken out, because fry are sensitive to sudden changes in the water chemistry. The fry should be fed several times a day. Many species need periodic sorting by size, so that larger fish do not cannibalize smaller fish. With favourable water conditions, regular water changes, and generous feeding, the fry should grow quickly. Cull unhealthy and deformed fish.

## Fishkeeping

***Figure:*** *An aquascaped freshwater aquarium*

Fishkeeping is a popular hobby concerned with keeping fish in a home aquarium or garden pond. There is also a fishkeeping industry, as a branch of agriculture.

## Types of Fishkeeping Systems

Fishkeepers are often known as "aquarists", since many of them are not solely interested in keeping fish. The hobby can be broadly divided into three specific disciplines according to the type of water the fish tolerate: freshwater, brackish, and marine (also called saltwater) fishkeeping.

### *Freshwater Aquarium*

A freshwater aquarium is a receptacle that holds one or more freshwater aquatic organisms for decorative, pet-keeping, or research purposes. Modern aquariums are most often made from transparent glass or acrylic glass. Typical inhabitants include fish, plants, amphibians, and invertebrates, such as snails and crustaceans.

***Figure:*** *Tropical freshwater aquarium*

Freshwater fish may be either coldwater or tropical species. Although freshwater aquariums can be set up as community tanks, coldwater and tropical fish are generally not mixed due to incompatibilities in temperature requirements. Coldwater aquariums house goldfish and other species that do not require a heating apparatus. Warmer temperatures would actually increase their metabolism and shorten their lifespan. For a tropical fish tank, maintaining a warm environmental temperature ranging between 75 to 80 °F (24 to 27 °C) enables the fish to thrive.

Aquariums may be decorated with sand or gravel, live or plastic plants, driftwood, rocks, and a variety of commercially made plastic sculptures. The smallest aquariums are fish bowls, but these are not recommended for most fish as they are generally too small.

***History:*** The earliest known aquariums were artificial fish ponds constructed by the ancient Sumerians over 4500 years ago. The ancient Assyrians, Egyptians, and Romans also kept fish in ponds for food and entertainment purposes. The ancient Chinese were the first culture to breed fish with any degree of success. They raised carp for food around 2000 B.C., and developed ornamental goldfish by selective breeding. Goldfish were introduced to Europe during the 18th century.

In the later 18th century, widespread public interest in the study of nature was awakening, and fish were kept in glass jars, porcelain containers, wooden tubs, and small artificial ponds. It was during this time that zoologist and botanist, Johann Matthaeus Bechstein, kept a large number of fishes and amphibians and laid down the foundation for aquarium and terrarium science. The concepts of the proper aquarium and terrarium were developed later by Nathaniel Bagshaw Ward in 1829. During the 19th century the idea of the "balanced aquarium" was developed. This approach was an attempt to mimic a balanced ecosystem in nature.

According to this method, fish waste could be consumed by plants, and plants along with the air surface of the water could supply oxygen for the fish. In 1869, the first tropical fish (the Paradise Fish) was imported from Asia. In these days, tropical tanks were kept warm by an open flame. Because early filters were noisy and expensive, fishkeeping was a hobby reserved for wealthy, scientifically inclined individuals.

***Figure:*** *A heavily stocked fancy goldfish aquarium*

In 1878, Rear-Admiral Daniel Ammon brought the first tropical fish from the Far East to the United States which lead to a decline in the popularity of goldfish. In the early 20th century, aeration, as well as particulate and charcoal filtration was introduced. The undergravel filter was introduced in the 1950s. By this time, the old idea of the balanced aquarium was viewed as unattainable and unnecessary by many people in the aquarium hobby, but it made a comeback at the end of the 20th Century with the rising popularity of the planted tank.

Today fishkeeping has become a popular hobby that almost anyone can do. Aquarium fish are both wild-caught and bred in Asia and Florida. Captive-bred species are inexpensive and widely available, and are less likely to be infected with diseases or parasites. Unfortunately, successive generations of inbred fish frequently have less colour and sport smaller fins than their wild counterparts.

### Fundamentals

A typical household freshwater aquarium set-up, apart from its aquatic tenants, consists of furnishings such as a gravel substrate, live or plastic plants, rocks, driftwood, a backcloth or background, and other decorations. Other equipment includes a canopy or hood as an aquarium cover, an aquarium stand or base, lighting accessories, a heater, a thermometre, air pumps, filtration apparatus, airstones, fish food, a fish net, water conditioner, water quality testing kits, a siphon hose or gravel cleaner, and a bucket for water changes.

Surface area and height are important in the set-up and maintenance of a living biotope. The surface area contributes to providing superior in-tank oxygenation and it also facilitates the creation of attractive aquatic themes. Freshwater environments benefit more from short and wide aquariums, due to the larger surface area they present to the air; this allows more oxygen to dissolve in the water, and the more oxygen there is, the more fish you can keep.

In general, a larger-sized aquarium provides a more stable water-world and the hobbyist can also acquire a greater number of fish. A large aquarium can also enhance aesthetic value. With regards to material, an all-glass aquarium is preferable due to its reasonable cost and its superior ability to resist scratches and discolouration. Indoor aquariums are normally placed far from windows, heating and cooling ducts of the house because direct sunlight and temperature changes can negatively affect the aquatic environment. Overexposure to sunlight leads to rapid algae growth inside and outside the tank. Sudden temperature variations are harmful to fish.

### Themes

Fish come in a large variety of species, from several different geographical regions. Most aquarium fish originated in Central

America, South America, Africa, or Asia. Fish can be kept in different combinations of species and in different kinds of aquatic environments. Four common themes include the community aquarium, the goldfish aquarium, the African cichlid aquarium, and the planted aquarium. A *community aquarium* refers to the mixing of fish and plants from different geographical areas with an emphasis on the colour and hardiness of the specimens. An example would be the combination of gouramis, tetras, and rasboras with a selection of hardy plants such as Hygrophila difformis, Hygrophila polysperma, and Vallisneria spiralis. Choosing fish that are peaceful and compatible with each other is important in a community tank.

A *goldfish aquarium* can be set up as a unfurnished and bare-bottom tank to emphasize the bright colouration of the fish. A combination of different varieties of goldfish and decorations that contrast with the vivid colours of the fish would make an attractive display. Live plants are not usually grown with goldfish, except for hardy, oxygenating plants like Egeria, because goldfish regularly disturb the substrate. They may also feed on softer-leaved plants. Plastic plants can be used instead.

An *African cichlid aquarium* commonly consists of Lake Tanganyika or Lake Malawi cichlid varieties, and generally requires a large number of rocks combined with a substrate of fine gravel or sand. The rocky environment should provide numerous caves and hiding places. Because cichlids, like goldfish, disturb the substrate by digging, plastic plants should be used as a substitute for live plants. However, real plants like Vallisneria or Anubias can be tried in a cichlid tank.

A *planted aquarium* emphasizes living plants as much as, or even more than fish. Large groupings of plant species such as Hygrophila, Limnophila, Rotala, Vallisneria, Echinodorus, and Cryptocorynes with a limited number of fish is a good example of a planted tank. It is important to select fish that will not damage the plants, such as small tetras, dwarf gouramis, cherry barbs, zebra danios, and White Clouds. Planted tanks may include $CO_2$ injection and a substrate fortified with laterite or, in the case of a low tech aquarium, a layer of potting soil under the gravel to provide nutrients for the plants.

A *biotope aquarium* is an aquarium that is designed to simulate a natural habitat, with the fish, plants, and furnishings all representative of a particular place in nature. Because only species that are found together in nature are allowed in a true biotope aquarium, these tanks are more challenging and less common than the other themes.

## Marine Aquarium

A marine aquarium is an aquarium that keeps marine plants and animals in a contained environment. Marine aquaria are further subdivided by hobbyists into fish only (FO), fish only with live rock (FOWLR), and reef aquaria. Fish only tanks often showcase large or aggressive marine fish species and generally rely on mechanical and chemical filtration. FOWLR and reef tanks use live rock, a material composed of coral skeletons harbouring beneficial nitrogen waste metabolising bacteria, as a means of more natural biological filtration.

Marine fishkeeping is different from its freshwater counterpart because of the fundamental differences in the constitution of saltwater and the resulting differences in the adaptation of its inhabitants. A stable marine aquarium requires more equipment than freshwater systems, and generally requires more stringent water quality monitoring. The inhabitants of a marine aquarium are often difficult to acquire and are usually more expensive than freshwater aquarium inhabitants. However, the inhabitants of saltwater aquariums are usually much more spectacular than freshwater aquarium fish.

***Marine Fishkeeping History:*** The first saltwater tanks were Venetian glass jars where the Romans kept anemones outdoors, but these systems were very short lived. The first personal saltwater fishkeeping began on a wider scale in the 1950s, starting with the basic rectangular glass aquariums (usually 20 gallon), still popular today. Bleached coral along with a substrate of coarse crushed coral were the norm. Algae, including beneficial types such as coralline algae, were viewed negatively and were generally removed. The clean, sterile tank was viewed as the healthiest. The Aztec Empire had 10 ponds of saltwater aquariums at Texcoco. During the early days of marine aquaria, saltwater was collected at local beaches. Natural saltwater contains many unwanted organisms and pollutants. Aquarium literature of the time suggests that the most commonly kept marine fish were the percula clownfish, sergeant major damselfish, small, brackish-water pufferfish and scats, jewelled blennies, and blue damsels.

Aquariums were equipped with large air compressors, and were heavily aerated and filtered (primarily with undergravel filters, a norm for some time). An ever-growing number of hobbyists experiencing the inconvenience of gathering natural sea water and the concurrent development of analytical chemistry techniques led to research into the chemical composition of sea water. Synthetic salt mixes were developed to replicate the chemical environment of the tropical ocean, including trace elements and salts. This advance made marine fishkeeping popular in areas without access to clean sea water.

Air driven, counter-current protein skimmers and reliable submersible electric heaters were invented in Germany. Various advances in filtration included trickle and hang-on filters, both allowing a more natural equilibrium in the aquarium environment.

The advancement of fluorescent lighting technologies to provide higher output, along with metal halide lighting, enabled the first reef tanks, making it possible to keep corals and other invertebrates without natural sunlight.

More efficient chemical testing allowed aquarists to understand the chemical properties of aquariums. By the 1980s, a biologically-based understanding of how to maintain an artificial ocean environment brought more successful and widespread marine fishkeeping.

### *Marine Aquarium Components*

***Figure:** Nano reef aquarium maintained at home*

The major components are an aquarium, usually made from glass or acrylic, filtration equipment, lighting, and an aquarium heater. Marine aquariums can range in volume from less than 80 litres, (<

20 US gal) to over 1,200 litres (300 US gal). Small volumes are more difficult to maintain due to the more rapid changes in water chemistry. The majority of saltwater aquariums are between 160 and 400 litres (40 and 100 US gal).

### Types of Marine Aquariums

Marine aquarists typically divide saltwater aquariums into those housing fish only, those housing fish with live rock, and those primarily designed to house corals and other invertebrates (also known as reef aquariums). Many fish hobbyists also divide the types of saltwater tanks based on the water temperatures at which they are kept.

### Tropical Marine

The most common type of saltwater fish tank, the tropical marine tank, houses marine animals from tropical climates. Usually kept between 24 to 28 °C (75 to 82 °F), these tanks include tropical reef tanks, as well as fish-only tanks. These tanks tend to have a low concentrations of microscopic plankton and other foods eaten by filter feeders. Most livestock for these aquariums are acquired through commercial means.

### Temperate Marine (Coldwater Marine)

***Figure:*** *A temperate marine aquarium*

One of the more obscure types of fish tanks, the coldwater marine tank, holds fish of temperate climates, with temperatures ranging around 10 to 24 °C (50 to 75 °F). While most of these tanks are not as aesthetically pleasing as their tropical counterparts, colourful species such as the ornate cowfish, blennies and sea anemones can be found. However most coldwater fish are gray or dull in colour.

Most colourful species are found in the western pacific and west coast of North America. These tanks also tend to require extra skill to maintain. A significant diversity of species exists. Since coldwater coral reefs only occur at great depths, most hobbyists are largely confined to fish, sea anemones, crustaceans, echinoderms, mollusks and Feather duster worms.

A few corals can be found at low depths. Since there are very few commercially available coldwater fish and invertebrates, hobbyists usually have to physically acquire specimens.

The most common way of doing this is by trolling or seining, learned hobbyists use the movement of tides and searching methods to find certain species. Unlike commercially available tropical fish, whose behaviour patterns and tank compatibilities are well documented, coldwater fish require much local ichthyology knowledge in order to maintain them. Many temperate fish have specific local diet requirements, while others, will eat just about any crustacean or frozen foods.

Some fish should not be kept with fish small enough to fit into its mouth, crabs or mollusks. Similarly some crabs can not be kept with some mollusks, while other fish, crabs, mollusks and echinoderms may be compatible with each-other. It takes experience before one can successfully gage the compatibility of the fish and invertebrates in one's area. Due to it being such a localised hobby, not many people go the route of local tanks. However, they are more cost-effective than reef tanks.

## Live Rock

Live rock is rock from the ocean that has been introduced into a saltwater aquarium. Along with live sand, it confers to the closed marine system multiple benefits desired by the saltwater aquarium hobbyist. The name sometimes leads to misunderstandings, as the "live rock" itself is not actually alive, but is made simply from the aragonite skeletons of long dead corals, or other calcareous organisms, which in the ocean form the majority of coral reefs. When taken from the ocean it is usually encrusted with coralline algae and inhabited by a multitude of marine organisms.

The many forms of micro and macroscopic marine life that live on and inside of the rock, which acts as an ideal habitat, give it the name "live rock".

***Figure:** Bleached coral skeletons, which can be inhabited by micro- and macro-organisms to form live rock*

## Origin

Live rock is harvested for use in the aquarium trade from collections in the wild near reefs, where parts may become detached from the main body of coral by storms. It may also be "seeded" from small coralline rocks by an aquaculturalist in warm ocean water, to be harvested later. Live rock can also be seeded by adding base rock to an active reef aquarium that already has live rock. Live rock harbours a wide variety of corals, algae, sponges, and other invertebrates, when they are collected. Corals added to the aquarium later will often become attached to the rock.

## Purpose

For the aquarium trade this rock is highly valued not only for the diversity of life it can bring to the closed marine environment, but its function as a superior biological filter that hosts both aerobic and anaerobic nitrifying bacteria required for the nitrogen cycle that processes waste. Live rock becomes the main biological nitrification base or biological filter of a saltwater aquarium. Additionally, live rocks have a stabilising effect on the water chemistry, in particular on helping to maintain constant pH by release of calcium carbonate. Lastly, live rock, especially when encrusted with multiple colours of coralline algae, becomes a major decorative element of the aquarium and provides shelter for the inhabitants. It is often used to build caves, arches, overhangs, or other structures in the tank, a practice known as aquascaping.

In J. Charles Delbeek's article *Your First Reef Aquarium*, he states,

> *"The use of live rock immediately introduces into the aquarium numerous algae, bacteria and small invertebrates all of which contribute to the overall quality of the aquarium water. Live rock has just as much, if not more, surface area for bacteria than a trickle filter. Since live rock in the aquarium contains various types of bacteria, algae and corals, waste products such as ammonia, nitrate and phosphate can have a number of fates. Ammonia, nitrate and phosphate are readily assimilated by algae and photosynthetic corals growing on and in the rock. Ammonia can also be quickly converted into nitrate by the bacteria on and in the rock. This nitrate can be either absorbed by the algae and corals, or it can be denitrified by bacteria in close proximity to the nitrate-producing bacteria."*

### Types

There are many different types of live rock. Each is named after the area from which it originated. A large amount of live rock comes from the Southern Pacific region, in areas such as Fiji, Tonga, and the Marshall Islands, as well as from the Caribbean. Each has its own distinct qualities that make it preferable to certain reef aquarists. For instance, live rock from the Fiji region is often porous and large, and rock from the Tonga region is often dense and elongated.

### Base Rock

Base rock is a generic term for aragonite rock that has no organisms growing in or on the rock. Base rock is often used as filler rock in the aquarium as it is much cheaper to purchase than live rock. In time, base rock will become colonised by living organisms.

Recently base rock that is mined from inland ancient reefs has become a popular way to keep the aquarium trade going sustainably. This rock is either maricultured and sold as live rock, or can be purchased and grown in the home aquarium.

Base rock can also be made from artificial rock called aragocrete, which is a hand made concrete from combining crushed aragonite, sand, and Portland cement.

After allowing the cement to dry, the pieces are sometimes acid washed to counteract the high pH of the materials, and then allowed to soak in clean water for one or more months. They generally tend to be heavier and less attractive when compared to natural base rock.

### Collection Ban

As of August 4, 2008 CITES (Convention on International Trade in Endangered Species) banned the collection of live rock from Tonga, the Marshall Islands, and the Cook Islands. This is due to the over-collecting of rock in these areas. The Secretariat has lifted all of the CITES restrictions in the recently banned countries for exporting species. This includes Tonga, Fiji, and Australia. Most notably, Tonga rock and corals and Fiji corals are available once again.

## Filter (Aquarium)

***Figure:*** *Air-driven corner filter*

Aquarium filters are critical components of both freshwater and marine aquaria. Aquarium filters remove physical and soluble chemical waste products from aquaria, simplifying maintenance. Furthermore, aquarium filters are necessary to support life as aquaria are relatively small, closed volumes of water compared to the natural environment of most fish. Animals, typically fish, kept in fish tanks produce waste from excrement and respiration. Another source of waste is uneaten food or plants and fish which have died.

These waste products collect in the tanks and contaminate the water. As the degree of contamination rises, the risk to the health of the aquaria increases and removal of the contamination becomes critical. Filtration is a common method used for maintenance of healthy aquaria.

### *Biological Filtration and the Nitrogen Cycle*

Proper management of the nitrogen cycle is a vital element of a successful aquarium. Excretia and other decomposing organic matter produce ammonia which is highly toxic to fish. Bacterial processes oxidise this ammonia into the slightly less toxic nitrites, and these are in turn oxidised to form the much less toxic nitrates. In the natural environment these nitrates are subsequently taken up by plants as fertilizer and this does indeed happen to some extent in an aquarium planted with real plants.

An aquarium is, however, an imperfect microcosm of the natural world. Aquariums are usually much more densely stocked with fish than the natural environment. This increases the amount of ammonia produced in the relatively small volume of the aquarium. The bacteria responsible for breaking down the ammonia colonise the surface of any objects inside the aquarium. In most case, a biological filter is nothing more than a chemically inert porous sponge, which provides a greatly enlarged surface area on which these bacteria can develop. These bacterial colonies take several weeks to form, during which time the aquarium is vulnerable to a condition commonly known as "new tank syndrome" if stocked with fish too quickly. Some systems incorporate bacteria capable of converting nitrates into nitrogen gas.

Accumulation of toxic ammonia from decomposing wastes is the largest cause of fish mortality in new, poorly maintained, or overloaded aquariums. In the artificial environment of the aquarium, the nitrogen cycle effectively ends with the production of nitrates. In order that the nitrate level does not build up to a harmful level regular partial water changes are required to remove the nitrates and introduce new, uncontaminated water.

## Mechanical and Chemical Filtration

The process of mechanical filtration removes particulate material from the water column. This particulate matter may include uneaten food, feces or plant or algal debris. Mechanical filtration is typically achieved by passing water through materials which act as a sieve, physically trapping the particulate matter. Removal of solid waste can be as simple as physical hand netting of debris, and/or involve highly complex equipment. All removal of solid wastes involve filtering water through some form of mesh in a process known as mechanical filtration.

The solid wastes are first collected, and then must be physically removed from the aquarium system. Mechanical filtration is ultimately ineffective if the solid wastes are not removed from the filter, and are allowed to decay and dissolve in the water.

Dissolved wastes are more difficult to remove from the water. Several techniques, collectively known as chemical filtration, are used for the removal of dissolved wastes, the most popular being the use of activated carbon and foam fractionation. To a certain extent, healthy plants extract dissolved chemical wastes from water when they grow, so plants can serve a role in the containment of dissolved wastes.

A final and less common situation requiring filtration involves the desire to sterilise water born pathogens. This sterilisation is accomplished by passing aquarium water through filtration devices which expose the water to high intensity ultraviolet light and/or exposing the water to dissolved ozone gas.

## Materials Suitable for Aquarium Filtration

Numerous materials are suitable as aquarium filtration media. These include synthetic wools, known in the aquarium hobby as filter wool, made of polyethylene terephthalate or nylon. Synthetic sponges or foams, various ceramic and sintered glass and silicon products along with igneous gravels are also used as mechanical filter materials. Materials with a greater surface area provide both mechanical and biological filtration. Some filter materials, such as plastic "bioballs", are best used for biological filtration.

With the notable exception of diatom filters, aquarium filters are rarely purely mechanical in action, as bacteria will colonise most filter materials effecting some degree of biological filtration. Activated carbon and zeolites are also frequently added to aquarium filters. These highly porous materials act as adsorbates binding various chemicals to their large external surfaces and also as sites of bacterial colonisation.

The simplest type of aquarium filter consists only of filter wool and activated carbon. The filter wool traps large debris and particles, and the activated carbon adsorbs smaller impurities. These should be changed regularly at suitable intervals. This is particularly important in the case of activated carbon filters, which may re-release their adsorbed contents in large (and therefore harmful) doses if they are allowed to saturate. Activated carbon adsorbs toxins on the surface of the carbon. It can be reactivated by boiling in clean water, and draining away the tainted water, and can then be re-used.

### Types of Aquarium Filters

***Figure:** A commercially available canister filter*

Numerous types of aquarium filters are commercially available, including:

### External Filters

External filters remove water from the aquarium which is then pushed (or pulled) through a series of different levels of filter media and returned to the aquarium. They are usually more effective and easier to maintain than internal filters.

### Canister Filters

Compared to filters that hang on the back of the aquarium, canister-style external filters offer a greater quantity of filter materials to be used along with a greater degree of flexibility with respect to filter material choice. Water enters the canister filled with the chosen filter material through an intake pipe at the bottom of the canister, passes through the material, and is fed back to the aquarium through the return pipe. Water is forced to circulate through the filter by a pump typically installed at the top of the canister. It is important to note that canister filters are sealed, fully flooded systems, meaning that the aquarium, intake pipe, filter interior and the return pipe form a continuous body of water. In this configuration both the intake and return path form two siphons, which precisely counterbalance each other. Under these circumstances, the filter pump does not have to spend any effort to lift the water back to the aquarium, regardless of how high the latter is installed above the canister. The pump should only be powerful enough to push the water through the filtering material as well as overcome the drag in the intake and return pipes. This makes canister filter pumps virtually insensitive to the height difference between the aquarium and the filter (although exceeding the manufacturer-specified height limit can lead to leaks).

Benefits of this type of filter are that they can provide a high volume of filter material without reducing the internal space in the aquarium, and that they can be disconnected from the tank for cleaning/ maintenance and replaced without disturbing the aquarium interior or occupants. Also, as a filter with external plumbing, it supports in-line installation of other aquarium equipment, such as water heaters and carbon dioxide diffusers. Such equipment can be removed from the tank and installed in-line into the return pipe of the filter. Disadvantages of canister filters include the increased cost and complexity relative to internal filters and difficulties in cleaning the tubes which transfer water to and from the aquarium. There is also the risk of a leak, which naturally is an issue for any filter placed outside of the aquarium.

Canister filters were initially designed to filter drinking water, under low pressure. Canister filters for aquariums use high water pressure, from a properly powered pump, to force water through the dense filter media. A pump can draw water from an under-gravel filter, and run it into a canister for double filtration.

### *Diatom Filters*

Diatom filters are used only for sporadic cleaning of tanks, they are not continuously operated on aquariums. These filters utilise diatomaceous earth to create an extremely fine filter down to 1 μm which removes particulate matter from the water column.

### *Trickle Filters*

Trickle filters, also known as wet/dry filters are another water filtration systems for marine and freshwater aquariums. This filter comes in two configurations, one which is placed on top of the aquarium (more rarely seen) and one which is placed below the aquarium (more common). If the wet/dry filter is placed on top of the aquarium, water is pumped over a number of perforated trays containing filter wool or some other filter material. The water trickles through the trays, keeping the filter wool wet but not completely submerged, allowing aerobic bacteria to grow and aiding biological filtration. The water returns to the aquarium like rain.

Alternatively, the wet/dry filter may be placed below the tank. In this design, water is fed by gravity to the filter below the aquarium. Prefiltered water is delivered to a perforated plate (drip plate). Prefiltering may take place in the aquarium via a foam block or sleeve in the overflow, or weir siphon, or it may be prefiltered by filter wool resting on the perforated plate.

The waste laden water from the aquarium spreads over the drip plate, and rains down through a medium. This may be a filter wool/ plastic grid rolled into a circular shape (DLS or "Double Layer Spiral") or any number of plastic media commonly known as Bio Balls. As the water cascades over the media, $CO_2$ is given off, oxygen is picked up, and bacteria convert the waste from the tank into less harmful materials. From here the water enters the sump. The sump may contain a number of compartments, each with its own filtration material. Often, heaters and thermostats are placed in the sump.

### *Fluidized Bed Filter*

The fluidized bed filter (FBF) is a biological reactor only. The principle is to direct water through a sand (or similar media) bed from below so that the sand becomes fluidized – behaves like a fluid. This mechanism is seen in liquifaction, quick sand, and industrial processes including municipal sewage treatment. The combined surface of all sand particles in the filter is very large, and so there is a large surface for aerobic denitrification bacteria. Therefore the size of the filter can be modest. The filter itself can be internal or external. In its simplest DIY internal version an FBF is very easy to build, with a container, sand, pump, and some plumbing. There are many variables: shape and size of the container, quantity of sand or equivalent, particle sizes, the pump's power, and plumbing.

***Figure:*** *Simple DIY fluidized sand bed filter*

### Algae Filters

Algae may be grown purposely, which removes chemicals from the water which need to be removed in order to have healthy fish, invertebrates and corals. This is a natural ("green") filtering method, which lets an aquarium operate the way oceans and lakes operate.

### Baffle Filters

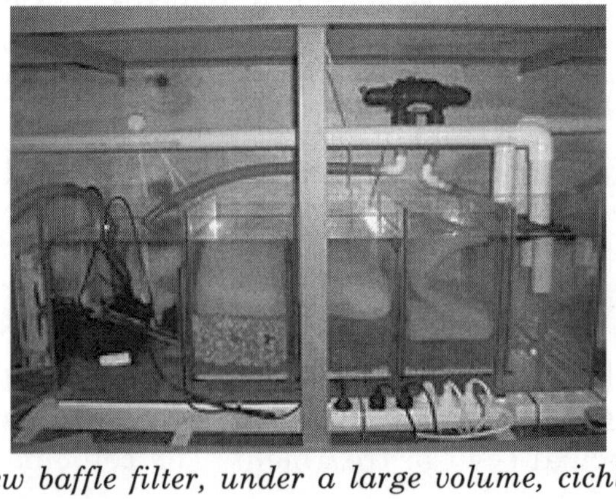

***Figure:*** *A new baffle filter, under a large volume, cichlid aquarium*

Baffle filters are similar to wet and dry, trickle filters in that they are generally situated below the aquarium. This type of filter consists of a series of baffles that the water must pass through in order to reach the pump which is returning water to the aquarium. These baffles then act much like a series of canister filters and can be filled with different filter media for different purposes.

### *Internal Filters*

Internal filters are, by definition, filters within the confines of the aquarium. These include the sponge filter, variations on the corner filter (pictured top right and left), foam cartridge filter and the undergravel filter. An internal filter may have an electric pump and thus be an internal power filter, often attached to the inside of aquaria via suction cups.

## Protein Skimmer

A protein skimmer or foam fractionator is a device used mostly in saltwater aquaria to remove organic compounds from the water before they break down into nitrogenous waste. Protein skimming is the only form of aquarium filtration that physically removes organic compounds before they begin to decompose, lightening the load on the biological filter and improving the water's redox potential. Although the process of foam fractionation is commonly known for removal of waste from aquaria, it is, in fact, a rapidly developing chemical process used in the large-scale removal of contaminants from wastewater streams and the enrichment of solutions of biomolecules.

### *Function*

Protein skimming removes certain organic compounds, including proteins and amino acids, by using the polarity of the protein itself. Due to their intrinsic charge, water-borne proteins are either repelled or attracted by the air/water interface and these molecules can be described as hydrophobic (such as fats or oils) or hydrophilic (such as salt, sugar, ammonia, most amino acids, and most inorganic compounds). However, some larger organic molecules can have both hydrophobic and hydrophilic portions.

These molecules are called amphipathic or amphiphilic. Commercial protein skimmers work by generating a large air/water interface, specifically by injecting large numbers of bubbles into the water column. In general, the smaller the bubbles the more effective the protein skimming is because the surface area of small bubbles occupying the same volume is much greater than the same volume of larger bubbles.

Large numbers of small bubbles present an enormous air/water interface for hydrophobic organic molecules and amphipathic organic molecules to collect on the bubble surface (the air/water interface). Water movement hastens diffusion of organic molecules, which effectively brings more organic molecules to the air/water interface and lets the organic molecules accumulate on the surface of the air bubbles. This process continues until the interface is saturated, unless the bubble is removed from the water or it bursts, in which case the accumulated molecules release back into the water column. However, it is important to note that further exposure of a saturated air bubble to organic molecules may continue to result in changes as compounds that bind more strongly may replace those molecules with a weaker binding that have already accumulated on the interface.

Although some aquarists believe that increasing the contact time (or dwell time as it is sometimes called) is always good, it is incorrect to claim that it is always better to increase the contact time between bubbles and the aquarium water. As the bubbles increase near the top of the protein skimmer water column, they become denser and the water begins to drain and create the foam that will carry the organic molecules to the skimmate collection cup or to a separate skimmate waste collector and the organic molecules, and any inorganic molecules that may have become bound to the organic molecules, will be exported from the water system.

In addition to the proteins removed by skimming, there are a number of other organic and inorganic molecules that are typically removed. These include a variety of fats, fatty acids, carbohydrates, metals such as copper and trace elements such as iodine. Particulates and other detritus is also removed, along with phytoplankton and bacteria.

This removal is highly desired by the aquarist and is often emphasized by the placement of the skimmer before other forms of filtration; lessening the burden of the filtration system as a whole. There is at least one published study that provides a detailed list of the export products found in protein skimmer skimmate. Protein skimmers are used to harvest algae and phytoplankton gently enough to maintain viability for culturing or commercial sale as live cultures.

### Design

All skimmers have key features in common: water flows through a chamber and is brought into contact with a column of fine bubbles. The bubbles collect proteins and other substances and carry them to the top of the device where the foam, but not the water, collects in a cup. Here the foam condenses to a liquid, which can be easily removed from the system. The material that collects in the cup can range from pale greenish-yellow, watery liquid to a thick black tar.

Consider this summary of optimal protein skimmer design by Randy Holmes-Farley:

For a skimmer to function maximally, the following things must take place:

1. A large amount of air/water interface must be generated.
2. Organic molecules must be allowed to collect at the air/water interface.
3. The bubbles forming this air/water interface must come together to form a foam.
4. The water in the foam must partially drain without the bubbles popping prematurely.
5. The drained foam must be separated from the bulk water and discarded.

Also under considerable recent attention has been the general shape of a skimmer as well. In particular, much attention has been given to the introduction of cone shaped skimmer units. Originally designed by Klaus Jensen in 2004, the concept was founded on the principle that a conical body allows the foam to accumulate more steadily through a gently sloping transition. This reduces the overall turbulence, resulting in more efficient skimming. While research into the specific benefits of the design are still being measured, early reviews of many conical skimmers have been positive overall.

Overall, protein skimmers can be classed in two ways depending on whether they operate by co-current flow or counter-current flow. In a co-current flow system, air is introduced at the bottom of the chamber and is in contact with the water as it rises upwards towards the collection chamber. In a counter-current system, air is forced into the system under pressure and moves against the flow of the water for a while before it rises up towards the collection cup. Because the air bubbles may be in contact with the water for a longer period in a counter-current flow system, protein skimmers of this type are considered by some to be more effective at removing organic wastes.

### Co-current Flow Systems

***Air Stone:*** The original method of protein skimming, running pressurised air through a diffuser to produce large quantities of micro bubbles, remains a viable, effective, and economic choice, although newer technologies may require lower maintenance. The air stone is most often an oblong, partially hollowed block of wood, most often of the genus Tilia. The most popular wooden air-stones for skimmers are made from limewood (Tilia europaea or European limewood) although basswood (Tilia americana or American Linden), works as well, may be cheaper and is often more readily available. The wooden blocks are drilled, tapped, fitted with an air fitting, and connected by air tubing to one or more air pumps delivering at least 1 cfm. The wooden air stone is placed at the bottom of a tall column of water. The tank water is pumped into the column, allowed to pass by the rising bubbles, and back into the tank. To get enough contact time with the bubble, these units can be many feet in height.

Air stone protein skimmers may be constructed as a DIY project from pvc pipes and fittings at low cost and with varying degrees of complexity.

While this method has been around for many years, many regard it as inefficient for larger systems or systems with large bio-loads.

### Venturi

The premise behind these skimmers is that a venturi pump, or aspirator, can be used to introduce the bubbles into the water stream. The tank water is pumped through the venturi, in which fine bubbles are introduced, then enters the skimmer body. This method was popular due to its compact size and high efficiency but venturi designs are now more likely to be included in other skimmer designs rather than as a simple venturi design.

# 7

# Models of Farming

Integrated fish farming can be recycled waste, full use of local resources, to maintain ecological balance, is a water, land, energy, grain, low effective breeding mode. Counties in Jiangxi Province of years now several models of integrated fish farming as outlined below: (a) fish - clam - livestock - poultry farming in three-dimensional surface based, fisheries, begun in the water fish, water pearl, underwater (Cage) raise mussels, pond duck or goose pig built homes, livestock and poultry manure fertilizer into the pond, conversion, or plankton, mussels provide natural food for fish.

Manure nutrients in a comprehensive, nitrogen, phosphorus, potassium and other nutrients were higher, and is mixed with a large number of livestock from the feces and undigested feed rippled feed. Therefore, pearl fish, livestock and poultry manure has a significant effect. Ratio of integrated fish farming, usually 400-450 per 0.067 ha fish pond tail, hanging about raising oyster 1000, and raised only a small mussels 3000-4000, with health pigs 7, 20 duck feather, with an annual fish, 150-200 kg, 2-3 kg of pearls. (B) of the fish - with integrated fish farming is generally used, the pool ridge, isolated land to grow forage terrestrial, terrestrial feed crop in the current, more common is the rotation ryegrass and Sudan grass.

Aquatic feeds such as water hyacinth and water lettuce. Terrestrial grasses, such as adequate water and fertilizer management, a full year of about 8000 kg, respectively, grass, according to the feed coefficient calculation, can produce about 450 kg of herbivorous fish. (C) Fish - Livestock - Agriculture with this model in pigs - grass - fish and vegetables - pigs - fish. Manure used as fertilizer to grow vegetables and grass. High production levels, each of 0.067 ha of grass productivity of phytoplankton primary production is 2-2.5 times, according to experiments, per 100 kg pig manure to feed the fish directly, filter-feeding and omnivorous fish can hyperplasia is 2.5 kg, but with in the grass, can produce 1.2 kg of terrestrial grass, can produce grass carp, bream 4.2 kg, and with the support filter, omnivorous fish, 1.5 kg, 5.7 kg total fish production, fish farming than pig manure fertilizer increased by 3.2 directly kg, an increase of 133%.

According to experts, the use of chicken fish, but also to prevent pond leakage, the average leakage reduction of 54% -76%, because the chicken can be uniformly deposited debris clogging the soil pool of bed spaces. By Fish - a chicken, fish - duck, fish - pork, fish - cattle, poultry, animal husbandry and integrated farming of fish, can prevent pond leakage. (D) fish - mulberry or fish - fruit combined use of ponds ridge sericulture, silkworm pupa, and pupa water fish, pond scum fat mulberry, hundreds of sand and pond scum that are fully utilised. Yela Tong time in trees, for planting rye grass or vegetables, used as fish feed, pond scum as a fertilizer.

It is estimated that 0.067 hectares per 2,000 kg of mulberry harvest mulberry leaves for silkworm silk could be 17.5 kg, 1200 kg of cocoon, and 38 kg 1 kg of silkworm excrement can be transformed into fish. In addition, each 0.067 ha planted in mulberry fields in winter rye grass or vegetables, generally can produce 3000 kg. Per 0.067 ha mulberry sericulture, silkworm production more than 130 kilograms, can be transformed into 1 kg of fish. (E) agricultural and sideline products processing and aquaculture industry with agricultural workers have meals category, grains, soy, wine, soy sauce and so on.

The fish processing industry, waste water sludge is a good pig feed. Pond ridge on the kinds of economic crops and oil crops, but also for the processing industry to provide a certain amount of raw materials. Bran and rice bran is the food processing industry fish waste water sludge is a good pig feed.

Pond ridge on the kinds of economic crops and oil crops, but also for the processing industry to provide a certain amount of raw materials. And rice bran is a byproduct of food processing industry, Lu, source wide, in addition to containing some protein, fat and more carbohydrates, but also rich in vitamin B, of great value as fish feed.

Lees is the residue from the brewing industry, the main nutrient content is less, but often contain more water-soluble vitamins and minerals, can be used as feed additives.

## Population Dynamics of Fisheries

A fishery is an area with an associated fish or aquatic population which is harvested for its commercial or recreational value. Fisheries can be wild or farmed. Population dynamics describes the ways in which a given population grows and shrinks over time, as controlled by birth, death, and emigration or immigration. It is the basis for understanding changing fishery patterns and issues such as habitat destruction, predation and optimal harvesting rates. The population dynamics of fisheries is used by fisheries scientists to determine sustainable yields.

The basic accounting relation for population dynamics is the BIDE model:

$$N_1 = N_0 + B " D + I " E$$

where $N_1$ is the number of individuals at time 1, $N_0$ is the number of individuals at time 0, *B*is the number of individuals born, *D* the number that died, *I* the number that immigrated, and *E* the number that emigrated between time 0 and time 1. While immigration and emigration can be present in wild fisheries, they are usually not measured.

A fishery population is affected by three dynamic rate functions:

- Birth rate or recruitment. Recruitment means reaching a certain size or reproductive stage. With fisheries, recruitment usually refers to the age a fish can be caught and counted in nets.
- Growth rate. This measures the growth of individuals in size and length. This is important in fisheries where the population is often measured in terms of biomass.
- Mortality. This includes harvest mortality and natural mortality. Natural mortality includes non-human predation, disease and old age.

If these rates are measured over different time intervals, the harvestable surplus of a fishery can be determined. The harvestable surplus is the number of individuals that can be harvested from the population without affecting long term stability (average population size). The harvest within the harvestable surplus is called compensatory mortality, where the harvest deaths are substituting for the deaths that would otherwise occur naturally. Harvest beyond that is additive mortality, harvest in addition to all the animals that would have died naturally. Care is needed when applying population dynamics to real world fisheries. Over-simplistic modelling of fisheries has resulted in the collapse of key

### History

The first principle of population dynamics is widely regarded as the exponential law of Malthus, as modelled by the Malthusian growth model. The early period was dominated by demographic studies such as the work of Benjamin Gompertz and Pierre François Verhulst in the early 19th century, who refined and adjusted the Malthusian demographic model. A more general model formulation was proposed by F.J. Richards in 1959, by which the models of Gompertz, Verhulst and also Ludwig von Bertalanffy are covered as special cases of the general formulation.

### Population Size

The population size (usually denoted by $N$) is the number of individual organisms in a population. The effective population size ($N_e$) was defined by Sewall Wright, who wrote two landmark papers on it (Wright 1931, 1938). He defined it as "the number of breeding individuals in an idealised population that would show the same amount of dispersion of allele frequencies under random genetic drift or the same amount of inbreeding as the population under consideration". It is a basic parameter in many models in population genetics. $N_e$ is usually less than $N$ (the absolute population size).

Small population size results in increased genetic drift. Population bottlenecks are when population size reduces for a short period of time. Overpopulation may indicate any case in which the population of any species of animal may exceed the carrying capacity of its ecological niche.

### Virtual Population Analysis

Virtual population analysis (VPA) is a modelling technique commonly used in fisheries science for reconstructing historical fish numbers using information on death of individuals each year. This death is usually partitioned into catch by fisheries and natural mortality. VPA is the most commonly used term to refer to cohort reconstruction techniques used in fisheries. It is virtual in the sense that the population size is not observed or measured directly but is inferred or back-calculated to have been a certain size in the past in order to support the observed fish catches and an assumed death rate owing to non-fishery related causes.

### Minimum Uiable Population

The minimum viable population (MVP) is a lower bound on the population of a species, such that it can survive in the wild. More specifically MVP is the smallest possible size at which a biological population can exist without facing extinction from natural disasters or demographic, environmental, or genetic stochasticity. The term "population" refers to the population of a species in the wild.

As a reference standard, MVP is usually given with a population survival probability of somewhere between ninety and ninety-five percent and calculated for between one hundred and one thousand years into the future.

The MVP can be calculated using computer simulations known as population viability analyses (PVA), where populations are modelled and future population dynamics are projected.

### Maximum Sustainable Yield

In population ecology and economics, the maximum sustainable yield or MSY is, theoretically, the largest catch that can be taken from a fishery stock over an indefinite period. Under the assumption of logistic growth, the MSY will be exactly at half the carrying capacity of a species, as this is the stage at when population growth is highest. The maximum sustainable yield is usually higher than the optimum sustainable yield.

This logistic model of growth is produced by a population introduced to a new habitat or with very poor numbers going through a lag phase of slow growth at first. Once it reaches a foothold population it will go through a rapid growth rate that will start to level off once the species approaches carrying capacity. The idea of maximum sustained yield is to decrease population density to the point of highest growth rate possible. This changes the number of the population, but the new number can be maintained indefinitely, ideally.

MSY is extensively used for fisheries management. Unlike the logistic (Schaefer) model, MSY in most modern fisheries models occurs at around 30% of the unexploited population size. This fraction differs among populations depending on the life history of the species and the age-specific selectivity of the fishing method.

However, the approach has been widely criticised as ignoring several key factors involved in fisheries management and has led to the devastating collapse of many fisheries. As a simple calculation, it ignores the size and age of the animal being taken, its reproductive status, and it focuses solely on the species in question, ignoring the damage to the ecosystem caused by the designated level of exploitation and the issue of by catch. Among conservation biologists it is widely regarded as dangerous and misused.

### *Recruitment*

Recruitment is the number of new young fish that enter a population in a given year. The size of fish populations can fluctuate by orders of magnitude over time, and five to 10 fold variations in abundance are usual. This variability applies across time spans ranging from a year to hundreds of years. Year to year fluctuations in the abundance of short lived forage fish can be nearly as great as the fluctuations that occur over decades or centuries. This suggests that fluctuations in reproductive and recruitment success are prime factors behind fluctuations in abundance. Annual fluctuations often seem random, and recruitment success often has a poor relationship to adult stock levels and fishing effort. This makes prediction difficult.

The recruitment problem is the problem of predicting the number of fish larvae in one season that will survive and become juvenile fish in the next season. It has been called "the central problem of fish population dynamics "and "the major problem in fisheries science". Fish produce huge volumes of larvae, but the volumes are very variable and mortality is high. This makes good predictions difficult.

According to Daniel Pauly, the definitive study was made in 1999 by Ransom Myers. Myers solved the problem "by assembling a large base of stock data and developing a complex mathematical model to sort it out. Out of that came the conclusion that a female in general produced three to five recruits per year for most fish."

### *Overfishing*

The notion of overfishing hinges on what is meant by an acceptable level of fishing.

A current operational model used by some fisheries for predicting acceptable levels is the Harvest Control Rule (HCR). This formalizes and summarizes a management strategy which can actively adapt to subsequent feedback. The HCR is a variable over which the management has some direct control and describes how the harvest is intended to be controlled by management in relation to the state of some indicator of stock status. For example, a harvest control rule can describe the various values of fishing mortality which will be aimed at for various values of the stock abundance. Constant catch and constant fishing mortality are two types of simple harvest control rules.

- Biological overfishing occurs when fishing mortality has reached a level where the stock biomass has negative marginal growth (slowing down biomass growth), as indicated by the red area in the figure. Fish are being taken out of the water so quickly that the replenishment of stock by breeding slows down. If the replenishment continues to slow down for long enough, replenishment will go into reverse and the population will decrease.
- Economic or bioeconomic overfishing additionally considers the cost of fishing and defines overfishing as a situation of negative marginal growth of resource rent. Fish are being taken out of the water so quickly that the growth in the profitability of fishing slows down. If this continues for long enough, profitability will decrease.

### Metapopulation

A metapopulation is a group of spatially separated populations of the same species which interact at some level. The term was coined by Richard Levins in 1969. The idea has been most broadly applied to species in naturally or artificially fragmented habitats. In Levins' own words, it consists of "a population of populations".

A metapopulation generally consists of several distinct populations together with areas of suitable habitat which are currently unoccupied. Each population cycles in relative independence of the other populations and eventually goes extinct as a consequence of demographic stochasticity (fluctuations in population size due to random demographic events); the smaller the population, the more prone it is to extinction.

Although individual populations have finite life-spans, the population as a whole is often stable because immigrants from one population (which may, for example, be experiencing a population boom) are likely to re-colonize habitat which has been left open by the extinction of another population. They may also emigrate to a small population and rescue that population from extinction (called the *rescue effect*).

### Age Class Structure

Age can be determined by counting growth rings in fish scales, otoliths, cross-sections of fin spines for species with thick spines such as triggerfish, or teeth for a few species. Each method has its merits and drawbacks. Fish scales are easiest to obtain, but may be unreliable if scales have fallen off of the fish and new ones grown in their places. Fin spines may be unreliable for the same reason, and most fish do not have spines of sufficient thickness for clear rings to be visible.

Otoliths will have stayed with the fish throughout its life history, but obtaining them requires killing the fish. Also, otoliths often require more preparation before ageing can occur. An age class structure with gaps in it, for instance a regular bell curve for the population of 1-5 year-old fish, excepting a very low population for the 3 year olds, implies a bad spawning year 3 years ago in that species. Often fish in younger age class structures have very low numbers because they were small enough to slip through the sampling nets, and may in fact have a very healthy population.

### Population Cycle

A population cycle occurs where populations rise and fall over a predictable period of time. There are some species where population numbers have reasonably predictable patterns of change although the full reasons for population cycles is one of the major unsolved ecological problems. There are a number of factors which influence population change such as availability of food, predators, diseases and climate.

### Trophic Cascades

Trophic cascades occur when predators in a food chain suppress the abundance of their prey, thereby releasing the next lower trophic level from predation (or herbivoury if the intermediate trophic level is an herbivore). For example, if the abundance of large piscivorous fish is increased in a lake, the abundance of their prey, zooplankti vorous fish, should decrease, large zooplankton abundance should increase, and phytoplankton biomass should decrease.

This theory has stimulated new research in many areas of ecology. Trophic cascades may also be important for understanding the effects of removing top predators from food webs, as humans have done in many places through hunting and fishing activities.

### Classic Examples

1. In lakes, piscivorous fish can dramatically reduce populations of zooplankti vorous fish, zooplankti vorous fish can dramatically alter fresh water zooplankton communities, and zooplankton grazing can in turn have large impacts on phytoplankton communities. Removal of piscivorous fish can change lake water from clear to green by allowing phytoplankton to flourish.
2. In the Eel River, in Northern California, fish (steel head and roach) consume fish larvae and predatory insects. These smaller predators prey on midge larvae, which feed on algae. Removal of the larger fish increases the abundance of algae.
3. In Pacific kelp forests, sea otters feed on sea urchins. In areas where sea otters have been hunted to extinction, sea urchins increase in abundance and decimate kelp.

A recent theory, the mesopredator release hypothesis, states that the decline of top predators in an ecosystem results in increased populations of medium-sized predators (mesopredators).

### Basic models

- The classic population equilibrium model is Verhulst's 1838 growth model:

$$\frac{nN}{dt} = rN\left(1 - \frac{N}{K}\right)$$

where $N(t)$ represents number of individuals at time $t$, $r$ the intrinsic growth rate and $K$ is the carrying capacity, or the maximum number of individuals that the environment can support.

- The individual growth model, published by von Bertalanffy in 1934, can be used to model the rate at which fish grow. It exists in a number of versions, but in its simplest form it is expressed as a differential equation of length ($L$) over time ($t$):

$$L'(t) = r_B(L\infty - L(L))$$

where $r_B$ is the von Bertalanffy growth rate and $L_\infty$ the ultimate length of the individual.

- Schaefer published a fishery equilibrium model based on the Verhulst model with an assumption of a bi-linear catch equation, often referred to as the Schaefer short-term catch equation:

$$H(E, X) = qEX$$

where the variables are; $H$, referring to catch (harvest) over a given period of time (e.g. a year); $E$, the fishing effort over the given period; $X$, the fish stock biomass at the beginning of the period (or the average biomass), and the parameter $q$ represents the catch ability of the stock.

Assuming the catch to equal the net natural growth in the population over the same period (X=0), the equilibrium catch is a function of the long term fishing effort $E$:

$$H(E) = qKE(1 - \frac{qE}{r})$$

$r$ and $K$ being biological parameters representing intrinsic growth rate and natural equilibrium biomass respectively.

§The Ricker model is a classic discrete population model which gives the expected number (or density) of individuals $N_{t+1}$ in generation $t + 1$ as a function of the number of individuals in the previous generation,

$$N_{t+1} = N_t e^{r(1 - \frac{Nt}{k})}$$

Here $r$ is interpreted as an intrinsic growth rate and $k$ as the carrying capacity of the environment. The Ricker model was introduced in the context of the fisheries by Ricker (1954).

- The Beverton Holt model, introduced in the context of fisheries in 1957, is a classic discrete-time population model which gives the expected number $n_{t+1}$ (or density) of individuals in generation $t + 1$ as a function of the number of individuals in the previous generation,

$$n_{t+1} = \frac{R_{o^n t}}{1 + n_t / M}$$

Here $R_0$ is interpreted as the proliferation rate per generation and $K = (R_0$ " 1) $M$ is the carrying capacity of the environment.

§Nurgaliev's law says:

$$\frac{dN}{dt} = aN^2 - bN$$

where $N$ is the size of a population, $a$ is a half of the average probability of a birth of a male (the same for females) of a potential arbitrary parents pair within a year, and $b$ is an average probability of a death of a fish within a year.

### *Predator-prey Equations*

The classic predator-prey equations are a pair of first order, non-linear, differential equations used to describe the dynamics of biological systems in which two species interact, one a predator and one its prey. They were proposed independently by Alfred J. Lotka in 1925 and Vito Volterra in 1926. An extension to these are the competitive Lotka-Volterra equations, which provide a simple model of the population dynamics of species competing for some common resource. In the 1930's Alexander Nicholson and Victor Bailey developed a model to describe the population dynamics of a coupled predator-prey system. The model assumes that predators search for prey at random, and that both predators and prey are assumed to be distributed in a non-contiguous ("clumped") fashion in the environment.

## Rice-Fish Farming Systems

The rice-fish farming project was launched in Guinée Forestière in 2000. It was based on a similar project a few years earlier in the central westerly area of Côte d'Ivoire. Both projects were furnanced by the Agence française de development. At the end of these eight years, some lessons can be drawn and these are outlined in the present report, which was written following a field assignment by one of its authors in October 2008, who had also been the author of the project ten years earlier.

### *Project Context*

***Conditions favourable to Fish Farming:*** Due to its geographical characteristics, Guinee Forestiere has good potential for the development of fish farming combined with agriculture. This long-standing forested area1 is abundantly and regularly rainfed for eight to ten months in the year. The lowlands cover approximately 15% of the agricultural area and are, for the most part, usable with a potential estimated at 3000 hectares (Nga Thot Ingénieurs conseils, 2007). The region's agro-climatic characteristics are well suited to fish farming because of the presence of permanently standing bodies of water in a large number of valleys, the water's warmth, the amount of sunshine it receives and the short dry seasons.

Over 600 km away from Conakry, the country's capital and principal port, Guinée Forestière is particularly landlocked. The road network is in poor condition and supplies of consumer goods, agricultural inputs and equipment are costly and uncertain. Public utilities are not very efficient and farm advisory teams are not sufficiently well trained to give satisfactory support to farmers, particularly as regards fish farming which is seen as a new type of activity. Population growth in Guinée Forestière is high. The population in the area is evaluated at over one million five hundred people, 78% of which live in rural areas.

The proximity of Sierra Leone and Liberia, where there were civil conflicts in the 90's, means that a high proportion – evaluated at 40% in 2001 – of the population would seem to be even now composed of refugees (CIRAD-HCR, 2001).

The region's demography has had a severe impact on the agricultural production systems which, traditionally, were based on hillside slash-and-burn rice cultivation. With land reserves running out, farmers were obliged to intensify and turned to hillside cash crops, coffee in particular, and to systematic use of lowlands for rice cultivation. Before the project started, there was practically no fish farming anywhere in the region. A few roughly excavated pools where farmers kept fish they had caught in the backwaters met with little success and mostly fell into disuse.

## Strong Demand for Animal Protein

Fish is a staple foodstuff in West Africa. It is sold frozen, dried or fresh. FAO evaluates at 22% the dietary protein contribution of fish in this part of Africa 2. In the poorest countries, this proportion sometimes exceeds 50%, in particular in areas where other animal products are rare and costly. In the West African coastal countries, the animal protein contribution of fish is high: over 60% in Guinea, Senegal, Ghana and Liberia.

In Guinea's forested region, protein deflicit is the foremost dietary deficiency affecting the poor. Bushmeat is increasingly hard to come by and beef (from Upper Guinea), chicken and pork (produced locally) are sold at twice or three times the price of fish in local markets. The consumption of fish, which is the main protein dietary contribution in the area, is estimated at 10 kg per head per year. It is below the national average of 22.5 kg/head/year.

The nutritional value of fish eaten in the region is mediocre. The frozen sea fish which makes up the major part of available resources (about 90%) (Oswald and Blanchet, 2001), arrives only after numerous breakdowns in the cold chain caused by road blocks in the rainy season and uncertain, defective or even inexistent supplies of fuel and electricity. Smoked fish deteriorates rapidly when ambient humidity is high. Fresh fish, however, is more often than not sold live, at the same price as frozen fish. In the circumstances, local consumers much prefer to buy fresh fish, so that from the start of the project there was a potential market for fresh fish which only needed to be developed.

### Pilot Project

***Issues at Stake:*** In a poor, rural, landlocked environment, bereft of any effective public support, the project's end purpose was fighting poverty, but there were two complementary objectives: frirstly diversifying farming activities to improve the level of monetary income; secondly, a sustainable increase of the local supply of both fish and rice 3.

The project strategy rested on the farmers' capacity for initiative and risk taking 4. As a counterpart, the project offered a technique for lowland fish farming, easily acquired because of locally acceptable investment levels. The technique was to be extended via a training project and support for locally recruited groups of volunteers.

There were, however, at the outset, a number of unknowns regarding in particular the degree of acceptance of this activity by farmers, the possibility of integrating it into existing farming systems and the magnitude of its economic, social and environmental effects. For these reasons, the project's initial objectives were limited: only 80 rice and fish farming units in some twenty villages were planned. It could therefore be viewed more as a pilot project aiming to demonstrate the feasibility of lowland rice-fish farming than as a full-scale project.

### *Grant Unancing*

In 1999, the project was granted € 1.8 million, donated to Guinea by the Agence française de developpement, for the 2000–2008 period. By the end of the project in June 20085, the grant was fully disbursed.

- However, due to non reimbursement by the Guinean government of its debt to AFD, disbursements stop(...)
- Not very expensive as mostly provided by volunteers.

Most of the costs involved were for technical assistance 6, leadership and training for fish farmers. On-site development was entirely at the expense of beneficiaries.

### *A Restricted Team to Help Farmers*

Building on the experience acquired in Cote d'Ivoire, the same project management and farmer support system for fish farmers was used. It involved AFVP (Association française des Volontaires du Progress), in partnership with APDRA-F (Association Pisciculture et Development Rural en Afrique – France). The team's composition was as follows: a Project Leader, an executive from the Fisheries Ministry (corresponding to the Guinean government's compensation, two volunteers and local team leaders. The project provided for the recruitment and training of these team leaders. Recruitment increased gradually in step with the growing number of candidates for the new activity, starting with two team leaders and rising to eight by the end of the project. A female team leader was also included in the team to provide support for the wives of fish farmers working on the family farm.

### *Promotion of an independent farming system*

As the owners of the land were the men, only a few women, widows generally, were able to participate in the activity. For leadership in the more remote areas, the project outsourced to a service provider, INADER7. The ponds are dug manually and gravity fed; therefore no pumping is required. Construction is limited to digging a dike perpendicular to the flow of water in the lowland. The barrier created by the dike retains water upstream covering between 0.2 and over 1 hectare, with an average surface of 0.3 hectares per production pond. A service pond for the production of alevins must also be constructed. A moulded concrete monk type drainage system is built so that the volume of water behind the dam can be entirely drained off and the fish farmer can regulate the water level, in particular for rice cultivation and to harvest all the fish once the pond is emptied. An overflow outlet is also needed to evacuate excess water in the event of flooding which is frequent and sometimes violent in that part of the world.

The average cost of constructing a pond is equivalent to about € 1000 per hectare. For a 30 are pond, the total cost of investment including a 30 metre dike downstream, an intermediate dike and a bypass channel with a 250 m3 earth flll, is the equivalent of 350 €, of which more than half is the cost of cement and tools and the remainder is for labour. 21 To minimise cost, fish farmers enlist the help of members of the family or of a mutual-help group in the village. Or else, jobbers can be employed, in which case the expense is added to the cost of building materials and tools and a sufficient amount of money must be saved up before starting. Construction, which is often spaced out in time depending on the state of the fish farmer's furnances, can take anywhere between two months and over a year. Some farmers stock their unfinished site with fish so that they can start production as quickly as possible, and use initial financial returns to complete construction.

## Farming the Pond

### *Polyculture*

Mixed fishery is the recommended practice, with as the main species, tilapia (Oreochromis niloticus), reared together with heterotis (Heterotis niloticus) and catfish caught in the wetlands (Heterobranchus isopterus). Fish weight at the end of the cycle should be 250–500 gr. for tilapia, 1.5–3 kg for heterotis and 400–600 gr for catfish. So as to control the very prolific tilapia population and limit the reproduction of wild fish, a small carnivorous fish, hemichromis (Hemichromis fasciatus), is generally added to the production pond.

### *Stocking the Ponds with Alevins*

So that the fish farmers can be independent, the project provides the initial generation of fish in each village or zone. Afterwards, the farmers themselves rear and produce the alevins in the hatchery ponds. In practice, although most farmers produce their own alevins, these are occasionally traded among villagers, which encourages the autonomy and sustainability of local production.

Sexing alevins, with the object of selecting the males, whose growth potential is higher, is done by farmers when they transfer the fish to the production pond.

### Extensive Breeding

At the outset, the possibility of adopting a semi-intensive breeding system was considered. But the idea was dropped almost immediately because a large working capital would have had to be mobilised to buy fish feed and farmers might have become over dependent on uncertain supply systems. Extensive breeding was therefore seen as more suitable for the local context in which the fish are fed free of charge by nutrients already present in the water. Furthermore, rice cultivation provides a natural source of enrichment through roots, stems and crop waste products. The farmers adapt fish density to existing environmental conditions. However, some intensification can be achieved if the ponds are fertilized with agricultural waste products (rice and maize bran) and animal husbandry waste (poultry droppings, liquid manure and cow dung, etc.).

### Fish Farming Yields

Production cycles last six months and in practice, farmers can achieve two cycles per annum. Once the system is fully operational, yields are approximately one tonne per hectare per year. There are, however, important variations in the early years, depending on how well the farmer adapts to this new technique and masters the water level management or on the availability of waste products to top up feed for the fish. Harvests are made up of approximately 60% tilapia, 30% heterotis and 10% of other species (catfish, hemichromis).

### Combining Rice Cultivation with Fishsh Farming

Generally, farmers produce one single annual rice growing cycle as against two fish crops. The main obstacle to a second annual rice growing season is a concentration of bird attacks on the rare grain crops available in the dry season. The losses are so high that they deter any new attempt at double cropping.

Rice yields per cycle total an average of 2 tonnes per hectare but can be as high as 4 tonnes. In contrast, in undeveloped wetlands, yields are never more than 1 tonne per hectare.

Rice cropping in developed lowlands involves a change in technique. The farmer empties his pond and replants seedlings immediately. Other work involved when growing crops in undeveloped lowland is no longer required: flooding obviates the need for clearing, the particularly strenuous work with the daba 8 is unnecessary. There is no need for weeding, except around the edges, because the pond eliminates weeds. Only harvesting is more difficult because it is done partly under water.

### *Agricultural calendar*

Combined rice and fish cultivation fit very well into the farm's production cycle and the agricultural calendar as there is no overlap with other activities. Bedding the rice seedlings and the rice harvest take place simultaneously for rice grown in the pond and fiooded rice grown in non developed wetlands. As for fish farming, dates for catching the fish are very flexible because production cycles can extend over six months or a year depending on each farmer's preferences and strategies regarding the size of the fish or need for income (feast days or beginning of school term, for example). Moreover, the sale and consumption of fish can be spread over the space of a month or two, since fish is stocked live in the service pond. This flexibility in the rice and fish production system and its efficient integration in other farming activities are assets which farmers appreciate.

## Organisation of the Activity: A Prerequisite for Successful Social Dynamics

### *Central Role of the Fish Farmer Groups*

The rice-fish farming activities sponsored by the project are only launched once a minimum of five candidates are identified in a village. This system of group support was chosen for three main reasons:

- grouping together, fish farmers form mutual-aid associations, facilitating the earthmoving work and thereby reducing investment costs and making more effective use of periods when farmers have time to spare;
- group leadership and training of farmers costs less and is more rewarding because members can benefit from discussion and interchange;
- the group helps the individuals concerned to gain more recognition in their village.

The project works as follows: once a month, a group leader stays a week in each village or zone.

Group activity begins with a planning session in which difficulties encountered and specific training needs can be evaluated.

In the course of the week, group and targeted training sessions are held. One of the system's strong points is that the group leaders stay long enough to strike up acquaintance with farmers, to hear of and understand their difficulties and build up trust.

Training courses in the group are designed for easy dissemination since they will later be passed on by members of the group to new candidates through local networking.

This method gives rise to supportive sponsorship provided by existing members to new farmers thus paving the way for acceptance of their help and advice.

For new candidates, the groups validates the application with an assessment of the technical and financial feasibility of the lowland development plan.

The candidate then studies and defines the project, with the help of members of his group and advice from the group leader, preliminary to launching it.

The group, assisted by the applicant's neighbours and the rural community, then verifies and validates the applicant's ownership of the land, so as to prevent any subsequent dispute.

Fish farmers' groups are in charge of training and also promoting and monitoring the services connected to the fish farming activity: manufacture of the monk drainage system, hiring of jobbers for earthmoving work.

This local development work is useful to optimise sharing out of the alevins in the various ponds and to reduce the cost of equipment and local labour.

As mentioned above, one of the project's main objectives is to make the fish farming groups autonomous. They must acquire the qualifications enabling their members to provide the important leadership, training and assistance services that make the fish farming project sustainable once the project has ended.

It is thanks to continuing support over time that the system can be fully effective.

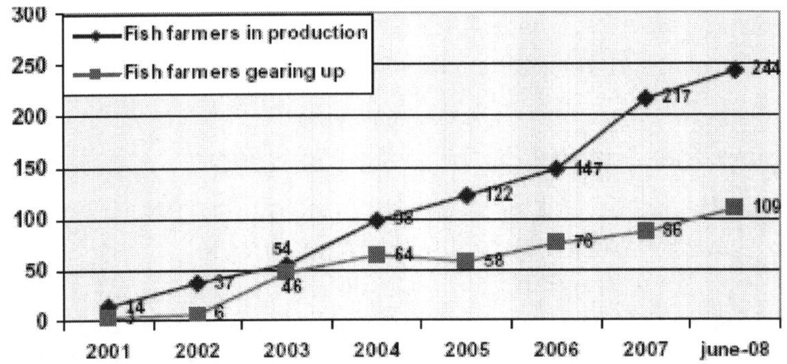


***Figure:*** *Number of fish farmers in guinee Forestiere*

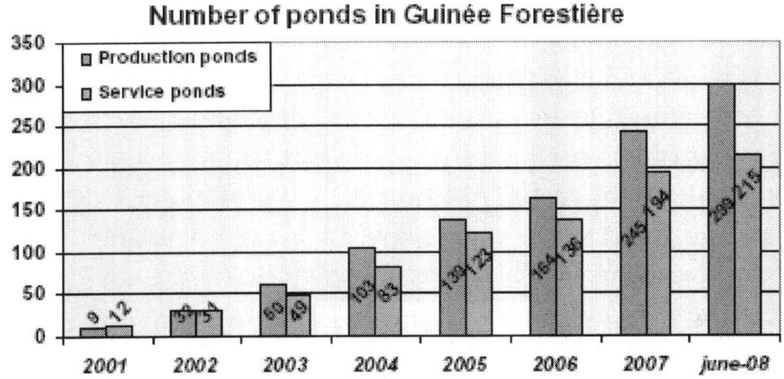


***Figure:*** *Number of ponds in Guinee Forestiere*

## Organisation of the Activity

In 2004, the groups created the Association des Pisciculteurs de Guinée Forestière (APGF) (Fish Farming Association in Guinée Forestière). The 350 fish-rice farmers of the area are members.

Through this association, fish farmers are in touch with national and foreign partners and can support the development of fish farming at a regional level. Contributions paid its members allow it to be supportive and representative.

It now needs to reinforce its internal organisation, in particular as regards rallying new candidates and become, as it intends to be, a true motor force for the development of this sector.

## Results and Positive Outcomes

Due to the project's innovative nature, start-up was slow. Farmers had to be persuaded to take the risk of branching out into this kind of production. Two years were spent on awareness-raising, but fish farming then started in earnest and initial expectations were more than met.

### *Encouraging the Spread of Rice-fish Farming*

On completion of the project, in June 2008, fish farming only prevailed in part of Guinée Forestière, mainly located in the N'Zerekore préfecture with an extension in Gueckedou. Out of the 350 fish farmers who were identified, 244 were already producing fish and 109 were gearing up. More than 500 ponds had been created over a 58 hectare area. Fish production is now 60 tonnes and rice paddy yields about 150 tonnes.

These results are encouraging. At the time of the 2008 mission, those in charge in the fish farmers association reported a large number – several hundred in fact – applications to set up new ponds and that this activity enjoyed a very favourable reputation in the villages. It seems reasonable to hope for a significant development of rice-fish farming in the next few years, thus substantiating the project's chosen intervention strategy and its financial dimensions.

### *On the Path to Food Security for the Local Population*

40 A third of the modest current production is self-consumed. The remaining two thirds are sold as fresh food in the home village and smaller quantities are sold further away. In October 2008, the price per kilo ranged from for tilapia and to for heterotis. The price of deep frozen fish is much the same as for tilapia. The product of fish farms is therefore competitive.

Live fish, when it is sold away from the home village, is generally transported on bicycles or mopeds in water-pulled cans so that they keep for several hours. In view of the quantities available for sale, this primitive system does not appear to be a limitation.

Altogether, fish production feeds 6000 people on the basis of a consumption figure of 10 kg per year. The number of fresh fish consumers would in fact be higher since frozen fish is still predominant. Therefore, as over 10 000 tonnes of frozen fish are sold annually in the region 9, the solvent demand for fresh fish is still largely unsatisfied. If the total 3000 hectares of lowland suitable for fish farming could be exploited, production would nonetheless only represent 25 to 30% of the quantity of frozen fish currently for sale.

## High Economic Profitability and Positive Social Impact

### *Profitability and Productivity of Farms*

The project is highly appreciated by farmers because, added to the net benefits procured by the fish farming activity, there is a significant improvement in profitability and productivity in rice cultivation.

Development of lowland to create a rice-fish farm gives a farmer the possibility of multiplying his total income per hectare by six (equivalent to 1620 €/ha) compared to the situation before development (280 €/ha), based on prudent evaluations of yields and prices. Two thirds of this income is derived from fish and one third from rice.

***Table:*** *Evaluation of average income generated per hectare with and without development.*

| | Units | Before the project | With the project |
|---|---|---|---|
| Rice yield | t/ha | 1 | 2 |
| Fish yield | t/ha | 0 | 1 |
| Unit price for rice | €/kg | 0.28 | 0.28 |
| Unit price for fish (weighted) | €/kg | 1.06 | 1.06 |
| Rice income | € | 280 | 560 |
| Fish income | € | 0 | 1060 |
| Total income* | € | 280 | 1620 |

* On-farm consumption included

Apart from the major contribution made by fish production, these higher rice-fish farming yields are due mainly to three causes: doubling (or even multiplying by as much as five in some cases) rice yields, a significant improvement in the productivity of rice-growing and greater flexibility in the agricultural calendar.

The improved yields are mainly due to soil fertility regeneration as a result of the lowland development work. Fish farming ponds are most often set up at the head of lowlands in rather infertile areas initially which are then emended by the deposit of silt and other organic matter. Moreover, the consequent constant presence of a sheet of water reduces the destructive effects on crops of pests such as caterpillars, rats and agouti.

With development, the amount of time that needs to be spent in the rice field is reduced by 30%. Whereas undeveloped lowland rice cultivation requires 160 days/ha, only 110 days are needed with developed lowland. The men are spared clearing and ploughing tasks and the women no longer need to do weeding.

In fish farming, production and marketing works on a flexible timetable so that income can be earned when needed, including during the dry season. Combined with the lighter burden of labour for rice production, farmers are better able to manage their work schedules and limit the use of outside labour. They can, for example, spend more time on their coffee plantation.

With this kind of profitability, time for recovery of investment costs can be as little as under two years, particularly when the major part of the work can be done by the farmer's family or by the mutual support group. Time for recovery is of course dependent on how soon the pond can become fully operational. Once the pond production has reached its full potential, return on investment can be as a little as a year.

Finally, by creating and distributing additional income at village level and increasing the food supply, rice-fish farming gives villagers the possibility of emerging from the poverty trap caused by the fact that the region is landlocked. This applies to both men and women. Nevertheless, rice fish farming involve only land owners.

## Effects on the Environment and on Health

### *Environmental Impact*

Despite a monitoring and evaluation system that still needs perfecting, the project has already provided some new in-sights regarding its effects on the environment. Obviously, lowland development has a traumatic effect on the natural environment. However, in these areas which are subject to growing anthropic pressures and to repeated stress (constant cropping, bush ores), creating permanently watered sites gives a large number of species an opportunity to and a niche into which they can fit.

Even though the developments only concern the bottom of hillsides and valleys, they are a partial response to the downgrading of the environment apparent in this zone: deforestation, evidence of reduced rainfall, rapid succession of slash-and-burn cycles with, as a consequence, the appearance of savanna types of landscapes. As we have noted, the developments improve soil fertility in the lowlands, help to regulate the flow of streams and rivers and runoffs from the drainage basin, replenish the groundwater table and reduce the level of hillside soil exploitation.

Water storage during the dry season (underground near the ponds and in the pond itself) and its availability for agricultural purposes has many advantages: changing savanna to post-forest formation, protecting crops against cure, etc. Nor is it impossible that much greater ecological complementarity exists between aquatic and hillside ecosystems than was previously thought. Whereas plateau cultivation is prone to phosphate deficiency, the pond ecosystem temporarily traps it, together with nitrogen.

Finally, it should be emphasized that the deliberately extensive production strategy prevents any chemical pollution of the water.

### Impact on Health

To date, there has been no outbreak of diseases following the construction and launching of the ponds. In particular, the schistosomiasis (bilharzia) vector seems to be eradicated by heterotis populations even when it is present in neighbouring rice paddies.

Mosquito larvae are eaten by tilapias. Pathogenic germs are eradicated because the ponds function like a constructed waste stabilisation system (in Côte d'Ivoire, contamination with salmonella when doubtful fertilisers were used, disappeared within two weeks). As ponds are systematically constructed so that they can be drained, when they are completely dry, pathogenic organisms including possible sources of disease which were contained in the water, die off and the local environmental conditions can be restored.

### Conclusions and Potential

Based on a pilot project fraught with uncertainty, the rice-fish farming activity in Guinée Forestière has made satisfactory progress and encouraged a large number of applications from potential operators. Should the number of constructions continue to grow or even accelerate, the food security and income improvement objectives could be fully achieved. The strategy used for the project, that is based on self-supporting groups and training of new applicants by experienced farmers, would be validated.

The main issue to be resolved would be the fish farmers' and their professional associations' capacity to cope with an extension of the policy. In Guinée Forestière, three conditions would seem to be required to consolidate existing achievements:

- reinforcement of the professional associations of rice-fish farmers so that they can carry the policy forward beyond the term set for the project;
- reinforcement of research, monitoring and evaluation, and regional exchange programmes;
- continuation of the preference for systems which are sparing in their use of agricultural inputs so as to maintain the principles of self-support and ownership by farmers.

The economic, social and environmental advantages of rice-fish farming have encouraged several donors to take an active interest in this type of operation.

A new project supported by IFAD has been identified to prolong the action of the Agence française de développement in Guinée Forestière. Similar projects are under study or under way in Liberia, Sierra Leone and Côte d'Ivoire. Coordination of aid in this field is emerging.

### *Rice-Fish Farming in Assam*

Rice fish farming can contribute to household income, contribute to food security and nutrition and contribute to improved sustainability of rice production. Generally two production systems have been recommended for culturing fish in the rice fields. They are, Simultaneous or Concurrent Method and Alternate or Rotational Method.

### *Selection Criteria of Rice Field*

- The field must hold water continuously for several months. The field should be hold water to a depth of 30 cm.
- The plot should be comparatively flat
- The land should be selected keeping in view that it will not be over flooded.
- The soil of the land should hold much water. Clay soil is better for this purpose and the soil pH should be around 6.5-7.5
- Those fields where water remains even after harvest of paddy

that is, there should be water, which is sufficient to continue fish culture.

- Fishes can be grown unaffected by chemicals or pesticides used to protect paddy.
- In land with slopes, a high dyke on the uphill side at the field is usually not needed. The layout of the land will help to confine the fish, if the rice field is basin shaped. This can save a lot of work because middle of the field is deepest portion and a very little effort is required to raise the dykes.
- The paddy field should have strong dykes to prevent leakage and to retain water up to desired level /depth.
- The plot should be at the close vicinity of the farmer's house so that better care can be taken up.

### Selection Criteria of Fish Species

- The fish should be capable of tolerating very shallow water level.
- Should withstand higher temperature (up to 40 OC) and variable temperature fluctuations
- Those species, which can withstand fairly high turbidity of water and poor oxygen concentration.
- The species, which have faster growth rate and should have desirable characteristics to grow to marketable size in short duration at the time of harvesting the rice.

### Selection of Rice Varieties

- The variety of paddy should be high yielding
- The variety should be highly disease resistant and susceptible to less attack from pests.
- Local paddy varieties of medium to long duration with non lodging characteristics are suitable.

### Preparation of Fields and Other Considerations

***Dyke construction:*** Embankments should have a height of 40-50 cm. Since water level for rice does not exceed 20 cm, such embankments will already have a free board of 20-30 cm. This is sufficient to prevent fish from jumping over.

***Provision of weirs and screens:*** Three types of screens can be provided to prevent escape of fish and to prevent entry of predatory fishes to the plot: e.g. bamboo slats, a basket, and a piece of fish net materials even a well perforated piece of sheet metal.

***Provisions of proper drains:*** The common practice is to temporarily breach a portion of embankment for water to get in or out and once the purpose is finished the breach portion be repaired. Bamboo tubes, hollowed out logs, metal pipes or bamboo chutes are also used. Depending upon the slope of the land three types of layout for construction of rice fish culture plot has been proposed :

- Perimeter trench model,
- Central pond model,
- Lateral trench model.

### *Perimeter Trench Model*

In this trenches are dug out in the periphery of the paddy field and the paddy cultivation area remains in the middle of the plot in zone of moderate elevation.

In a plot of 1 ha, the area in the central part of the field for paddy cultivation is about 0.67 ha. The perimeter trenches may occupy about 0.2 ha and perimeter dykes may occupy another 0.12 ha. Design and construction is that the trench is about 6 meter at the top, 3.5 m at the base and depth is 1.2 m. The perimeter may be 1 m in wide at the crest and 3 m at the base.

### *Central Pond Model*

In this case fish culture area remains at the centre or at the middle of the plot and the paddy growing areas lies surrounding the pond. In a plot of 1 ha area, part of the field for paddy cultivation is about 0.65 ha and the area for the fishpond is 0.35 ha with peripheral dyke space of 0.002 ha.

The dimensions of perimeter dyke may be 20cm in width, base is of 50 cm width and height is of 30 cm.

### *Lateral Trench Model*

Trapezoidal trenches are dug at the sides of paddy plot of dimensions top width 18 m, base 15m and depth is of 1.5 m. There is also to be a peripheral dyke like other two previous cases.

In a plot of 1 ha area; part of the field for paddy cultivation is about 0.61 ha and the area for the fishpond is 0.27ha with peripheral dyke space of 0.12 ha.

### Water Management

Continuous flooding up to the maximum tolerated by rice without affecting its rice production is recommended and it is generally 15-20 cm. At that depth, the effective water depth of 65-70 cm is available to the fish in refuge. This is sufficient to provide the fishes a cooler place when the shallow water over the rice warms up to as high as 40OC. The increased depth means a greater volume of water for rice –fish culture.

### Fertilization

Application of fertilizers increases the nutrient concentration in water, which flourish the growth of phytoplankton, which are directly consumed by the fish or support good growth of zooplankton. Compost fertilizer for deep-water variety @30 tons per ha can be applied. In organic fertilizers for deep water variety Paddy as Nitrogen @120kg/ ha, P205 @ 60 kg/ha and K20 @ 60 kg/ha can be applied.

### Stocking Patters

Rice fish culture may involve the stocking of young fry for the production of fingerlings (nursery operation) or the growing of fingerlings to marketable size (growing operation). Rice fish farming may either be the culture of only one species (monoculture) or a combination of two or more species of fish and crustaceans (polyculture). Generally, stocking density of fish depends on its size, species and the fertility of the land.

It is best to wait until the rice is well established before releasing fish seed. Fish can be stocked once two or three tillers have appeared for which the usual waiting is 1- 3 weeks after transplanting or 4-6 weeks after seeding.

### Feeding

To boost the growth of fish, rice bran and mustard oil cake in the ratio of 70:30 @2-3% of the body weight of fishes can be provided.

## Biodiversity Benefits: Rice-Fish Farming in China

China has revived a 2000-year-old tradition of raising fish in rice fields, boosting yields and farm incomes while preserving biodiversity, reducing agrochemicals use, and even beating back malaria. Tell an Englishman he could get both his fish and his chips from the same bit of land and he'd say you were barking mad.

***Figure:*** *Farmers plant rice seedlings on the outskirts of Yingtan, in China's Jiangxi province. Many farmers are also raising fish in the some fields. (Source: Reuters)*

But that's what the Chinese and other Asians have done for centuries: it may be fish and rice not fish and chips, but this marriage of agriculture and aquaculture is an all-too rare biodiversity success story. Rice-fish farming "epitomizes an ecosystems approach" which contributes to "eco-environmental benefits and sustainable development" says the UN Food and Agriculture Organisation (FAO) in a 2009 report by researcher Miao Weimin.

### *A Match Made in Heaven*

Heaven on earth to the ancient Chinese was a place that provided plenty of rice and fish. Ever-ingenious, they began cultivating these fundamentals of the Chinese diet together, putting baby carp into their paddies as they planted rice seedlings. Come harvest time, the fish were ready for the pot. Rice-fish farming in China can be traced back over 2000 years.

But in the 1960s and 1970s the Green Revolution swept through Asia, bringing with it more intensive rice farming using large amounts of fertilizers, herbicides and pesticides, all toxic to fish. Rice-fish farming collapsed. The government revived it in the 1980s and 1990s, however, and the rice-fish farm area increased from 648,000 hectares in 1985 to 1.55 million hectares in 2007 while fish production leapt from 81,700 tons to 1.16 million tons, according to the FAO. Today China gets more fish products from fish farming than from the sea, and rice-fish farming provides over one quarter of those fish. There's more variety too. Traditional species like carp and tilapia have been joined by high-value species such as swamp eel, crawfish and freshwater shrimp.

### Rice-Fish Farming From Scratch

Farmers dig trenches along the edges and across the rice field. They use the excavated soil to raise the banks around the rice field higher than normal paddies so that no water or fish can escape. These fortified banks can support other crops, such as soybean, rapeseed or grasses which can be used as feed for the fish. A popular feed is azolla—an aquatic fern that grows very rapidly in watery soils—which also doubles up as an organic nitrogen fertilizer.

The trenches are a refuge for the fish during the rice planting and harvesting seasons or when water levels are low, but they also keep the fish away from the rice seedlings when they are first planted. At the start of the rice growing cycle the field is flooded. Farmers plant their rice and put juvenile fish into the trenches.

Once the rice is established the fish are let into the rice fields where they grow until harvest time, when the water is drained and the fish collected. These rice-fish culture system can produce from 300 to 900 kilograms of fish per hectare or 300 to 750 kilograms of prawns or crabs per hectare.

### Don't the Fish Eat the Rice?

To stop them nibbling on the young rice plants, farmers feed the fish with green grasses and other feeds. But there are plenty more sources of fish food in the paddies: ones that the farmers are glad to get rid of. Fish eat pests such as stemborer and leaffolder and so they reduce the need for pesticides.

They also eat weeds that choke rice paddies and bacteria such as sheath blight disease and disease-infected leaves, thus reducing herbicide use.

Furthermore, by gobbling up all this biomass the fish are reducing methane emissions from decomposing vegetation by up to 30 percent compared to conventional rice farming. And of course all this free food on offer means that the farmer has to spend less time and money growing or buying supplementary feed for the fish. Having eaten, the fish excrete rich organic fertilizer, which reduces the need for petrochemical-based fertilizers.

The fish don't just protect rice health, but human health too. Fish eat mosquito larvae, and a study of one village in Guangxi province showed that as rice-fish cultivation increased malaria cases decreased.

### *But does it Make Money?*

The Communist Party backed rice-fish farming to make farmers richer. Beijing wanted to increase rural prosperity to keep people on the land and stem the destabilizing tide of urban migrants.

According to the FAO, it worked. "The income of some 2–3 million rural households has been significantly improved through rice-fish culture in China," says Miao Weimin, who says the net income from rice-fish culture is 2,000 to 4,000 dollars per hectare.

That's not just because of the extra income from more high-value fish—crustaceans and other high-value species fetch 3 to 5 times the price of carp. Rice yields have improved as well, even though the space taken up by trenches for the fish can be up to 15 percent of the rice field. Fish farmed rice is generally high quality rice too: organic rice that attracts a premium from discerning Chinese consumers. The farmer's income can be increased by 2 to 4 times compared with sole crop farming.

What was a subsistence rice-fish culture has transformed into a green/organic food production system that enhances biodiversity and conserves a symbiotic rice-fish ecosystem. Rice-fish farming faces challenges, however.

It relies on a good quality water supply and China is notorious for polluted waterways. China also faces a drought-prone future, particularly in the Yangtze River basin where much rice-fish farming is done.

And not many rice-fish farmers are following the environmentally friendly best practices described above. Better and more widespread education and training is required, as well as more stringent organic certification schemes.

That's especially true if, as seems likely, rice-fish farming continues expanding. Currently, only 15 percent of the available area in China is used for rice-fish culture. Looking further afield, the potential in the rest of Asia is enormous, with about 140 million hectares—nearly 90 percent of the world total—under rice cultivation. The Philippines, Bangladesh, India and Thailand are among the other Asian countries reviving or exploring rice-fish farming.

Their success or failure could have huge consequences for future food security, poverty alleviation, biodiversity preservation and human health.

## Animal-Fish Farming Systems

A pond on a small farm can mean much better all-round farm integration and greater overall production. Apart from their use for ducks and geese, ponds can raise a useful supply of fish. A framework for the integration of poultry and fish production in the Tropics and Sub-Tropics is proposed. Linkages between poultry and fish production may bring benefits to both components and the wider farming system. Both poultry production and processing wastes have value as nutrient inputs to fish and the water used for fish culture can act for evaparative cooling of poultry and fertigation of crops.

The conceptual basis of controlled eutrophication of fishponds using poultry manure for the production of herbivorous fish is compared to feeding of poultry abbatoir wastes to carnivorous fish. A comparison of poultry production systems in terms of their potential for integration of fish culture is made; the modern feedlot is compared and contrasted with traditional systems. The nature of poultry wastes is reviewed with respect to the effect of poultry strain/species, diet, and poultry and waste management. The impacts of use of bedding materials, frequency of waste collection and contaminants are discussed.

The use of poultry feedlot waste alone for fish culture is compared to the use of waste and additional fertilisers or feeds. The relative value of wastes from scavenging poultry alone or together with other inputs is analysed. The political economy of current poultry and fish production are considered in this article. The impacts on public health and the environment are also considered in the paper. Fish raised in semi-intensive, freshwater systems provide the major proportion of farmed, global production (FAO, 1995). A high proportion of this aquaculture occurs in rapidly developing Asia, which is also experiencing sharply increased consumption of poultry. Semi-intensive systems are usually based on ponds fertilized with livestock manure and fed with low cost supplementary feeds.

The integration of livestock and fish can increase overall production intensity and economise on land, labour and water requirements for both poultry and fish.

For example, one hectare of static water fish ponds can 'process' the wastes of up to 1500 poultry, producing fish in quantities of up to 10 MT/ha without other feeds or fertilisers. Also, since effluents are few, environmental impacts are minimal.

The importance of poultry wastes in aquaculture is relatively recent. In areas of traditional fish culture, ruminant and pig manure have predominated as pond fertilisers in the Subcontinent and China respectively. Poultry manure was not used to any extent probably because small flock size and extensive management precluded collection. Livestock production systems, and opportunities for reuse of wastes and by-products, are changing. Vertical integration of the poultry industry by agribusiness has been stimulated by the biology and widespread acceptability of poultry, particularly chickens. Global trends in livestock production indicate that poultry, particularly layer and broiler chickens, are increasing faster than any other (FAO,1989,1990,1991,1993).

The intensive nature of modern poultry production and processing tends to concentrate high quality by-products, and this has stimulated their reuse. A range of poultry by-products are produced and reused in livestock feeds including feather meal, bloodmeal, poultry litter meal etc. (Möller, 1980), and poultry wastes are also used as fertilisers and soil conditioners. Economic growth is fuelling demand for both poultry and fish in many parts of the Asia Pacific region and a major question is the extent to which their integration should be promoted further here and elsewhere.

***Table:*** *Matrix of livestock waste qualities and suitability for use in aquaculture. (*** = high to * = low)*

*Factors increasing relative suitability for aquaculture*

| *Livestocktype* | *Collectability* | *Acceptability* | *Nutrient density* | *Low opportunity cost* | *lack of deleterious compounds* |
|---|---|---|---|---|---|
| Poultry | | | | | |
| Feedlot | *** | *** | *** | * | *** |
| Scavenging | * | ** | ** | ** | ** |
| Pigs | | | | | |
| Feedlot | *** | * | ** | ** | *** |
| Scavenging | * | * | * | ** | ** |
| Ruminants | | | | | |
| Feedlot | *** | ** | ** | ** | ** |
| Scavenging | * | ** | * | ** | * |

Poultry production wastes have inherent qualities that make them particularly valuable for fish production compared to other livestock wastes. Poultry wastes are more nutrient dense than other livestock wastes.

Typically they contain less moisture, fibre and compounds such as tannins that discolour water when used as fish pond fertilisers. Commercial 'feedlot' production leads to concentration of nutrient-rich waste which can be handled and transported cost-effectively. This may mean that the opportunity costs of poultry manure are higher, and their use for horticulture particularly is common. The small individual size of poultry also allows their confinement and production directly over fish ponds.

Poultry manure has been used widely in both fresh and brackish water aquaculture. In the latter, Penaeid shrimp, Milkfish (Channos channos) and Tilapia (Oreochromis sp.) have been the principle species raised. Inland culture systems in which poultry and fish such as the carps, tilapias and catfish are raised in commercial and subsistence sytems are the focus of this review.

Poultry manure is now widely used in commercial freshwater aquaculture. In central Thailand, use of livestock wastes is the norm in the production of cheaper herbivorous fish. In other areas of Asia, intensification of culture using high quality feeds has reduced the importance of poultry waste to fish production. Predisposing factors to intensification include shortage of land or water and high product prices, but ready availability and competitively priced quality feeds are also critical. Wohlfarth and Schroeder (1979) identified the relative price of feeds and manures as being critical to determining input strategies.

Most published data concern integration of fish culture with modern poultry systems which are typically inappropriate for resource-poor farmers. Village or backyard poultry systems predominate in areas where modern breeds and systems are absent, or coexist in competition with them. Recent research indicates that integration of such poultry and backyard fish culture can also bring benefits at little extra cost.

### *Waste-fed Aquaculture*

Much of the nutrient content of feed given to poultry is voided as excretory or faecal waste. These nutrients can be used to support fish culture by their action as fertilisers that stimulate production of natural food organisms, such as phytoplankton, and detritus. A variety of carps and tilapias can grow rapidly on such natural feeds alone.

Stable and high water temperatures and sunlight ensure year-round growth of both fish and their natural feeds.

The tropics, in which average water temperatures remain above 25 degree C, are ideal for culturing fish using poultry waste as inputs, although it is also practised in subtropical and sub-temperate climates during suitable periods of the year (>20 degree C).

Poultry wastes and by-products can provide the feed support of aquaculture across a range of intensities. Poultry wastes may act mainly (1) indirectly or (2) directly to support fish production.

Poultry manure can be used fresh, or after processing, to enhance natural food production in sun-lit tropical ponds (Figure 2). Although some nutrition may be derived directly from the waste, natural feed produced on the nutrients released from the wastes is more important. Fish feeding low in the food web - the carps and tilapias benefit most from this type of management since they can utilise plankton, benthic and detrital food organisms effectively.

Several factors affect the level of waste loading and standing stock of fish that can be supported. Greater sensitivity to dissolved oxygen limits carps to standing stocks of <3 MT/ha whereas tilapias may be harvested at standing stocks of over 5 MT/ha. Water quality, particularly the level of dissolved oxygen early morning, therefore limit the amount of wastes that can used. Input levels in excess of 75 kg DM/ha/d typically 'overload' the system over a typical fish culture cycle (4-8 months), causing early morning deficits of oxygen. Balancing the production of wastes from poultry and the requirement of the fishpond is a key aspect of management.

The quality of poultry wastes used in fish culture varies greatly. High levels of spilt feed, for example, increase direct feeding value. Nutrient composition may be a useful guide to value but the availability or release of nutrients to the food web may be more important.

Conventional feed ingredients have been 'replaced' with dried poultry wastes of various types, but low metabolisable energy and digestible protein levels limit their usefulness (Wohlfarth and Schroeder, 1979).

Poultry processing by-products such as chicken bones, intestines and whole carcasses have greater value as 'direct' feeds and are normally used for higher value fish species raised more intensively. High fish standing stocks can be maintained and yields produced using this type of product and management. Processing wastes can be used fresh, or after further processing, as good quality supplementary, or complete, feeds.

### Traditional Aquaculture

A lack of nutrients was a major constraint to traditional aquaculture and this remains true for much of the fish culture practised in the developing World. Yields from carp-based polycultures in China were limited until recently by the paucity of diets for pigs and grass carp (Ctenopharyngodon idella), and their manures, which provided a large portion of the nutrients entering the food web (Ruddle et al. 1983 ; Guo and Bradshaw, 1993). Recycling and reuse of nutrients on-farm has a long tradition born of necessity in the denser-settled areas of Asia. However, the high outputs of fish and other products from integrated systems reported from China and elsewhere in recent decades are related to greater nutrient inputs from off-farm.

Lack of nutrients and sub-optimal stock management remain major constraints to the production of fish on small-scale farms, as they are for traditional livestock management generally. Greater outputs of fish necessitate more nutrient inputs to be used than are available on typical resource-poor farms. Such inputs may be either direct fish feeds or, feeds for livestock that in turn produce waste used in fish culture. Both need to be purchased from off-farm to supplement better reuse of on-farm wastes.

### Factors Affecting Use of Poultry Wastes in Fish Culture

The type of poultry production system can greatly influence the amount of fish produced. Poultry systems producing nutrient-rich and collectable wastes are most valuable for fish production. A broad dichotomy exists between 'modern', normally intensive poultry production and 'traditional', extensive systems and this affects potential for integration with fish culture (Little, 1995). Edwards et al. (1983) describing the level of integration of poultry with fish in Central Thailand found large differences between small and large producers. In contrast with pigs in which even small herds are typically integrated, poultry (chickens, ducks) flock sizes of less than 100 birds were unlikely to be cultured with fish but larger flocks (>400) were usually integrated. Modern 'feedlots' raise large flocks and are generally capital intensive, highly dependent on off-farm support and profit-orientated. Generally raised on optimal, processed feeds in 'feedlots', production cycles are rapid and all the high quality waste can be collected for use in fish culture.

In contrast, Klausner (1966) observing traditional management in a Northeast Thai village said that 'the owners feel that there is not much point in taking pains and expending money in caring for chickens, when chickens seem quite capable of caring for themselves'.

Many factors appear to constrain close integration of traditional poultry and fish culture. The poor quality supplementary feeds usually given and the fact that confinement is restricted to overnight result in less, and poorer quality, manure being available for use in fish culture. Moreover, farm households may already be using the poultry waste which is collectable for other purposes such as fertilizing backyard crops. Recent analysis of current poultry production in small-scale farming households reveals a marginal but important niche.

### Poultry Production Waste Characteristics

Poultry manures are nutrient-rich, but there is great variability in their quality at the time of use as fish production inputs. Although between 72-79 % of the dietary nitrogen, 61-87% phosphorous and 82-92% of the potassium was present in feedlot, egg-laying hens (Taiganides, 1978), the variability in terms of nutrients available (g nutrient/bird/day) can be much greater. The impacts of the gradual improvements in food conversion efficiencies attained by modern breeds and feeds are probably overidden by other factors, especially diet. Poultry raised on a balanced ration produce a higher quality, more nutrient dense waste than those fed a supplementary feed.

Species, size and sex of bird directly affect the quantity of manure produced. The amount of feed spilt during feeding and drinking also varies with these factors together with the nature of the feed and feeding practice.

Generally, larger birds produce more waste than small; the waste production increases rapidly over the rearing period of modern broilers as a result. Layers produce more calcium and phosphorous-rich excreta than broilers and the waste of replacement birds fed restricted diets high in fibre is correspondingly poorer than laying birds.

In scavenging systems, manure quality is greatly affected by the quality and quantity of supplementary feeds, which in turn affects fish production. Egg-laying ducks fed paddy grain at night produced poorer quality manure than those fed rice bran.

The amount of nitrogen and phosphorous in the manure was 50% and 25% respectively of that found in ducks fed relatively nutrient dense village rice bran. Restricted feeding of rice bran during night-time confinement to muscovy ducks (Cairhina moschata) scavenging during the day reduced both quantity and quality of collectable wastes. Nitrogen in wastes declined with the level of restricted feed given from 1.28 g N/duck/day, for birds fed ad libitum to 0.55 g N/duck/day for ducks restricted to 50% of ad libitum feeding levels. (AFE, 1992).

Supplementary feeds of different types drastically affect waste characteristics and their value for fish culture. In a trial in which three different supplementary feeds (village rice bran, ground maize and ground sorghum) were fed to pekin and muscovy ducks, both waste quantity and quality was affected.

The degree of wastage, related to palatability and physical attributes of the feed, was an important factor (see below) but the intake and proximate composition greatly affected the value of waste for fish culture (Niang, 1990). Manures derived from maize-fed ducks were high in nitrogen, sorghum were intermediate and rice bran low reflecting the composition of the feeds themselves. Total nutrients in the waste tended to be higher than in the feeds suggesting the scavenged food tended to be of higher food value than the supplement.

Spilled feed is a loss to the poultry system but a gain to the fish because of it's direct feeding value. The method of food presentation (timing, frequency, location) affects the amount of feed available directly for fish. Feedlot ducks fed complete diets appear to waste less than birds allowed to scavenge during the day and given access to supplementary feed at night.

Feed processing can reduce spillage; up to 15 % of granulated feeds may be lost compared to 10% if the same duck feed is pelletised (Barash et al.,1982). Feeding behaviour and the nature of different feeds may increase the amounts of feed available directly for fish. Waste feed left in the waterer comprised more than 25% of the collectable dry matter from scavenging muscovy ducks fed a supplement of village rice bran (AASP, 1996).

Poultry species, strain and environment affect the normal conditions of poultry management in tropical environments and these interact to determine the final characteristics of wastes available for fish culture.

The density of birds in a given system and their method of confinement -in small cages or batteries (such as for chicken layers) or in pens with bedding material (litter) affects the management of both the poultry and their waste. Confinement directly over fish ponds is used for both broiler and layer chickens but pens that give access to ponds stocked with fish for drinking and/or bathing are generally used only for ducks and geese.

Poultry house litter (PHL), which can be broiler, replacement or layer bird waste is produced from poultry raised in houses with bedding materials of various types.

The type and management of these materials can affect nutrient content and availability for fish culture. Fermentation can result in losses of nitrogen, volatalised as ammonia, or becoming refractory and unavailable.

Some vitamins may increase (vitamin B12) and some antibiotics (e.g. Chortetracycline) decrease with duration of storage (Muller, 1980).

***Table:*** *Check list of factors affecting characteristics of poultry waste and its use for aquaculture (modified after Mš ller, 1980).*

- used fresh or collected, stored and transported (CST)
- nature of bedding materials (bulk density, particle size, moisture retention capacity, compressibility, penetrability, hygroscopicty, biodegradabilty)
- type of bird (size, growth rate, efficiency, sex)
- housing (open, closed)
- litter management (regular/irregular removal)
- nature of ingredients in poultry ration (digestibility, nutrient density and composition)
- type of storage (aerobic, anaerobic, exposure to temperature, rain, wind)
- quantity of bedding materials per surface unit (nutrient dilution, microorganism activity).

***Table:** Mean Calcium and Phosphorus Content (on DM)*

| *Type of Waste* | *Ca* | *P* | *Ratio P:Ca* |
|---|---|---|---|
| Broiler manure* | 1.9 | 1.7 | 1 : 1.1 |
| Broiler litter (one batch) | 1.6 | 1.4 | 1 : 1.1 |
| Replacement bird manure* | 2.3 | 2.1 | 1 : 1.1 |
| Layer manure* | 7.5 | 2.6 | 1 : 2.9 |
| Cattle manure* | 1.3 | 0.8 | 1 : 1.6 |
| Pig manure** | 3.5 | 2.6 | 1 : 1.3 |

Sources: * Muller, 1974-75, ** Pearce, 1977 cited by Muller 1980

## Action of Wastes in the Pond

The rate of nitrogen and phosphorous release, particularly in the most available forms, (dissolved inorganic nitrogen, DIN; soluble reactive phosphorous, SRP) has been used as an indicator of wastes value for fertilization of fish ponds. Laboratory leaching experiments indicated that DIN was rapidly released as ammonia-N , levelling off at 6 mg NH4-N/g DM chicken manure after 4-5 days (Knud-Hansen et al., 1991). Storage of duck wastes under aerobic conditions for a period of 4 weeks reduced both total nitrogen in the waste and the amount released subsequently as ammonia (Ullah, 1989).

The type of ingredients fed poultry can affect the subsequent manure quality and release of nutrients for pond fertilization. Subsitution of a cassava mixture of leaf and root meal for village rice bran in complete diets of broiler muscovy ducks resulted in a more nitrogen-rich manure but a similar cumulative release of nitrogen (5.5-6.7 mg/g DM).

Release of DIN varied between 20-74% of the total in the waste (Figure 8). A greater proportion of phosphorous was released as SRP in all the wastes and the amount was inversely related to the level of cassava in the diet (AIT data).

Manure obtained from scavenging muscovy ducks fed variable levels of a rice bran supplement (100, 75 and 50% of ad libitum) had different release characteristics. Significantly more DIN was released by ducks fed less supplementary rice bran suggesting that the protein in the natural feeds ingested during scavenging were less refractory than the nitrogen contained in village rice bran. SRP showed the inverse trend, with ducks fed ad libitum producing manure richer in phosphorous, of which more was released in the available form (AIT data).

Manures release other factors apart from nutrients that may have adverse effects on water quality and inevitably, fish production. Shevgoor et al. (1994) found that tannins and flavanoids were a major factor in the poor water quality observed in ruminant manure-fed systems.

Substitution of cassava leaf for rice bran in complete diets for muscovy ducks correlated with increased levels of tannin released from manure (AIT data). The value of manures, including poultry manure, as a source of detritus and the role of detritus in the direct nutrition of fish has been much debated (Schroeder, 1978; Colman and Edwards, 1987).

The stimulation of bacterial production, both in the water column and sediments is known to be stimulated by addition of poultry waste (Moriarty, 1987). Animals that filter feed or graze on bacteria attached to detritus directly, or consume the grazers, are therefore likely to benefit directly through this mechanism. Both food and dissolved oxygen are required to maintain high fish yields and phytoplankton, both alive and as detritus, is the most important source of both in fertilized ponds.

### *Classification of Poultry-fish Systems*

A framework for poultry-fish systems. Feedlot and scavenging poultry represent two ends of a continuum of systems (Little and Edwards, 1994).

The type of producer and characteristics of the production system and waste collection methods are distinct. In both cases however, poultry wastes may be part of a range of inputs used to produce fish. The use of poultry processing wastes is distinct and considered seperately, although this strategy is linked closely to feedlot broiler production.

Van der Lingen described the concept of increased carrying capacity and fish yields if more nutritional inputs are complemented with higher stocking densities (Edwards, 1986) and (Figure 10). Yields from fertilization alone may be increased with the use of supplementary feeds. Further increases in density and yield rely on improvements in feed quality and quantity so that they become the primary source of nutrition to the fish. Poultry wastes are used across a range of intensities, and for different purposes. Poultry wastes, inorganic fertilisers and feeds are to some extent substitutable. Poultry waste can be used in place of inorganics or feeds, inorganics in place of manures or feeds, and feeds in place of either type of fertiliser. Thus if manures are in short supply, inorganics can be used to optimise nutrient loadings and feeds to further increase yields. Feed may be substituted, to some extent, with fertilisers.

### *Feedlot Systems*

Most of the poultry-fish systems described in the literature use waste from feedlots. Modern breeds of poultry raised on balanced feeds give the most nutrient-rich waste and produce the most fish, but systems are frequently suboptimal resulting in inefficient waste or space use. Poultry manure is used either directly on-site, through the siting of poultry houses over ponds, or after collection, storage and transport to the site of fish culture.

Construction of the poultry house over the pond allows waste to drop directly in, saving labour costs. Also, in the periurban, flood-prone land often used, the cost to fill land for poultry housing, and the opportunity cost of land itself, are reduced. Confining poultry next to, or over water can also improve their productivity under tropical conditions. Evaparative cooling can reduce heat stress in broilers (Theimsiri, 1992) and access to water improves feather quality of ducks, although growth may suffer (Edwards, 1986). Ducks free ranging over ponds in large numbers can damage dykes and cause water quality problems, restriction of the ducks to the water and pen prevents this problem (Edwards et al., 1983). However, evidence from on-station research and farmers suggests that access to complete feeds and some degree of scavenging optimises egg production in Khaki-Campbell ducks (AIT, 1986).

Fish species is a critical factor in determining loading rates of poultry waste since there is a range of sensitivity to dissolved oxygen among the commonly cultured fish species. Air-breathing fish, such as clarias catfish, and the silver-striped catfish, Pangasius hypothalmus can tolerate the highest input levels and, at the high stocking densities normally raised, also require extra feed to sustain growth. These fish species are also probably inefficient at using the phytoplankton-dominated food web. In contrast, the microphagous Nile tilapia is more sensitive to low dissolved oxygen early morning but thrives at numbers of poultry between 1000 and 1500 egg-laying ducks/ha without other inputs. Using poultry manure alone net extrapolated yields of up to 12 MT or a standing stocks of 5-6 MT/ha appear possible in monocultures of tilapia.

Polycultures of carps are often considered most efficient in waste-fed ponds but greater sensitivity to dissolved oxygen necessitates lower input levels. Hopkins and Cruz (1982) found poor survival of the more sensitive common carp in a tilapia-dominated polyculture, and Yakupitiyage et al. (1991) recorded poor survival of large silver barb (Puntius gonionotus) under similar circumstances. Research using more sensitive Indian Major carps has normally been undertaken at lower loading rates.

Temperature regime affects the level of wastes that can be used in ponds and this is reflected in the lower stocking densities in eastern Europe (Edwards, 1986). Low temperatures reduce the amount of waste that can be processed by a given area of fish ponds. The fish kills reported in Hong Kong (Sin, 1980) are related to a continued build up of waste during the cool season causing a subsequent bacterial and plankton boom as temperatures rise. This phenomeneon, equivalent to a massive overloading, quickly removes oxygen from the water.

The dynamics of poultry flocks can make management of the waste for fish culture problematic. Direct use of egg-laying poultry for instance, in which the birds are of constant weight and produce fairly constant levels of waste, are easier to manage than broilers in which waste availability is cyclical (Hopkins and Cruz, 1982) ; (Figure 12). The timely availability of replacement stock, vetinary support, and market demand may be critical to maintaining both poultry, and their waste, production.

Higher loadings of waste necessitate water exchange or mechanical aeration to maintain dissolved oxygen. Green et al. (1994). significantly improved yields of Nile tilapia at high loading rates of chicken manure (1000 kg DM/ha/week) using mechanical aeration to ensure high survival of fish.

Additional aeration at levels of 10% saturation were sufficient to improve yields by 20% over unaerated ponds. Regular exchange of water to reduce phytoplankton biomass can alleviate water quality problems caused by overloading. In a well designed system this would be avoided as effluents reflect inefficent nutrient reuse and could cause negative impacts on surrounding environment. Overloading of poultry waste can also be avoided by housing poultry over concrete or earthen floors rather than directly over ponds and regular manual or mechanical collection and addition. This option may reduce construction costs considerably and also enables farmers to sell manure surplus to their requirements.

### *Supplementation of Feedlot Wastes*

Various factors may limit the size of a poultry flock that a farmer can manage and integrate with fish culture, reducing wastes to levels below optimum. Edwards et al.(1983) found that problems marketing duck eggs, and high feed costs, constrained small-scale farmers maintaining even 30 ducks in feedlots over small ponds (200m2) and suggested the waste from smaller numbers of ducks could be supplemented with direct feeding of fish.

Farmers with limited numbers of poultry for their pond area need additional nutrient inputs to optimise productivity and fish yield and inorganic fertilisers may be the most cost-effective option. Inorganix fertilisers are a cheaper form of nitrogen and phosphorous than purchased poultry manure in many situations (Table 6), and highest yields of Nile Tilapia were achieved with relatively low loadings of poultry manure.

The optimal level of poultry manure in ponds fertilized with high levels of inorganic fertilisers (5 kg N/ha/d) was found to be around 75 kg/ha/week for a monoculture of Nile tilapia in Thailand (Knud-Hansen et al.,1991). These low levels reflect the subtle balance of dissolved oxygen and food production in a highly eutrophic pond. Green et al. (1994) recorded similarly high yields (>20 kg/ha/day) of Nile tilapia using higher levels of chicken manure in combination with inorganic fertilisers. There are indicators, however, that compared to tilapia, carps raised at lower nutrient loading rates perform better when fertilized with organic manures in addition to inorganics (AASP, 1996). Also, the use inorganics may be constrained by their poor availability or high opportunity cost.

Supplementing use of poultry manure with cheap and available direct feeding of fish is an alternative strategy. The impact of supplementary feed on yields of fish in ponds fertilized with poultry waste is affected by many factors.

The level of natural feed to some extent affects the effectiveness of th supplementary feed; more natural feed allows greater feeding of a high-energy supplement to 'spare' the protein requirements and support the growth of more fish (Hepher, 1981). Optimising levels of supplementary feeds is complicated by the variable levels of waste poultry feed mixed with the manure.

Strategic use of supplementary feeds such as rice bran can boosted yields of a duck manure-fed Nile tilapia monoculture by between 10-150% (AIT, 1986).

The use of feeds in this way, however, may not be cost effective. One trial clearly demonstrated the 'Law of Diminishing returns' when a low feeding rate (1% Bodywt/day) increased yields by between 28-40% profitably, depending on duck manure level. A further doubling in feeding rate (2% Bodywt/day) increased yields further by a mere 4% or reduced them by 16% respectively.

***Table:*** *Economic comparison of different fertilizers with respect to available nitrogen (N), phosphorus (P), and carbon (C) (US$ = 25 Baht), (Knud-Hansen, 1993)*

| *Fertilizer* | *Cost (baht/50 g)* | *Available N (baht/kg)* | *Available P (baht/kg)* | *Available C (baht/kg)* |
|---|---|---|---|---|
| Chicken manure | 40 A | 76[b] | 194[c] | 7[d] |
| Urea | 240 | 10 | - | 24 |
| TSP | 450 | - | 45 | - |
| $NaHCO_3$ | 1000 | - | - | 140 |

a. Wet weight

b. Assumes 40 % dry weight of total N is available (Knud-Hansen et al. 1991).

c. Assumes 10 % dry weight of total P is available (Knud-Hansen et al. 1993).

d. Assumes 50 % dry weight of organic C oxidizes to DIC.

This suggests that overall dry matter loadings into ponds receiving both feeds and fertilisers should be considered. The variable response of fish species with different feeding niches within the polyculture also illustrates that supplementary feeding should be strategic. Although manure level and rice bran acted independently to boost overrall yields, Nile tilapia responded most to duck density and silver barb only to feeding rate.

Supplementary feeding of fertilized ponds is only necessary if the carrying capacity is exceeded. Green et al. (1994) found no benefits to yields in poultry-manure fertilized ponds when high quality pelleted feed was also given. Probably, growth could be supported by natural feed alone at the low fish stocking densities (2/m2) used.

Yields may also be constrained by other factors such as seasonally low temperature and dissolved oxygen levels. The relatively low yields reported by in Hong Kong (1.5-4.7 MT/ha/year; Sin, 1980), despite supplementary feeding of carp polycultures fertilized with duck manure, appear to be related to such water quality factors. Green et al. (1994) also reported poorer yields in supplementary-fed, poultry waste-fertilized ponds during cooler periods. Reducing feeding costs of more intensive systems by fertilizing ponds with poultry manure is another strategy that has attracted attention by farmers and researchers alike. Clearly, the fish species raised need to be suitable for culture in plankton-rich, waste-fed systems. Green et al. (1994) found that the tambaqui (Colossoma macropomum), in contrast to the Nile tilapia, grew poorly in fertilized systems without supplementary feed.

Manuring and feeding appears to be a cost effective strategy for intensification of Nile tilapia culture. In Taiwan yields of up to 18 MT/ha are produced in duck manure-fertilized, aerated ponds in which the fish are also fed pellets in which the fish are also fed pellets on demand (Liao and Chen 1983).

Fertilization is also essential component of high-yielding, commercial polycultures in Isreal (Hepher and Pruginin's 1981). Feed costs may also be reduced by feeding only in the later stages of the culture period, when the nutritional needs of the fish exceed the level supported by natural feed alone.

Green et al. (1994) found that at densities of 1 fish/m2, tilapia could be raised on poultry waste alone for the first 3 months of an 137 day production without any differences in final yield. Fattening with pelleted feed over the final 50 days increased yields and individual fish size significantly.

### Scavenging Systems

The use of scavenging poultry wastes in aquaculture is rare; few systems have been described in anything but qualitative terms in the literature. The inherent variability of low input scavenging systems, including waste output, is much greater than conventional poultry-fish systems; the production function between waste level and fish production, for example, is far more variable.

The relationship between number of poultry and fishpond area is less clear cut when the wastes from scavenging birds are used. Waste collection is normally limited to overnight to allow enough time for the poultry to obtain natural foods, but absolute amounts of waste collected may still be high. If supplementary feeds are given ad libitum, and all wastes are used for fish culture (see above), dry matter levels/ bird may be higher than those produced in feedlot systems. Overloading of these wastes can have clear negative impacts on water quality and fish yields.

Developing fish culture based solely or partly on the wastes of current poultry production requires an understanding of feed constraints. In Northeast Thailand, the main supplementary feeds used for poultry, village rice bran and unmilled paddy rice, are available only in limited quantities. Feeding restricted amounts of feeds to a larger flock of poultry can result in more poultry and fish from the same amount of rice bran (AASP, 1996), provided that the scavenging feed resource base us significant (Tadelle,1997). The quality of scavenging environment might be expected to affect the requirement for supplementary feed and the final quality of wastes produced. Natural forage is frequently seasonal and crop harvests may produce short-lived abundances of residues (dropped paddy, spilt maize) that will affect waste composition.

The relatively low nutrient density of wastes from scavenging poultry fed supplementary feeds explains the rationale for using them as partial inputs into fish culture. Farmers understand this limitation; in a study of farmer practice in Udorn Thani, farmers tended to use a variety of inputs including plant leaves, ruminant manure and rice bran in addition to poultry manure (AASP, 1996). The quality and quantity of supplementary fed to scavenging poultry are key factors in determinging waste characteristics (see above) and subsequent fish yield. Feeds high in cassava products generally depressed fish yields, possibly due to the levels of tannins and unavailability of nutrients. Rice bran, corn and sorghum-fed ducks produced highly dissimilar wastes and subsequent fish yields based on similar numbers of ducks were vary variable (Naing, 1990). Egg-laying ducks allowed to scavenge, and fed either supplementary rice bran or paddy rice at night showed that tradeoffs may be involved. Egg yields were higher, but fish yields barely half that from ponds in which ducks were fed rice bran.

Inorganic fertilization can have a major impact on yields (x 100%) of microphagous fish such as Nile tilapia in ponds receiving scavenging poultry wastes. In small ponds, the relative amounts required are also affordable, given the value of the fish produced (Edwards et al, 1996).

***Table:*** *Restricted feeding of scavenging ducks makes better use of ther same amount of rice bran.*

| ***For Every 100 kg rice bran*** |
|---|
| • 14.0 Kg duck + 5.0 - 6.8 Kg fish if ducks fed *ad libitum* (100 %) |
| • 17.0 Kg duck + 7.3 - 10.2 Kg fish if ducks fed restricted (75 %) |
| • 20.3 Kg duck + 9.6 - 10.8 Kg fish if ducks fed restricted (50 %) |

(The range of fish yields reflects the duck stocking density/pond mean; lower stocking density/larger pond gives higher yield)

### *Poultry Processing Wastes*

Boneless chicken meat is now an international commodity that resource-rich developing countries, with vertically integrated poultry industries, compete to produce and market. Low labour and feed costs, and good infrastructure, are necessary preconditions to develop the business that can produce large amounts of high quality by-products suitable for intensive fish culture. Poultry slaughterhouse wastes are in great demand for feeding hybrid clarias catfish (Clarias macrocephalus x Clarias garipinus) in Thailand. Heads, viscera and thigh bones are the main by-products fed fresh after simple on-farm grinding and mixing with a binder. Food conversion ratios of 4-5 (wet:wet basis) are attained under farm conditions, similar to levels reported by Prinsloo and Schoonbee (1987) for the use of chicken offal and dead whole chickens fed to Clarias gariepinus.

### *Benefits -the Political Economy of Poultry-fish Systems*

Modern agribusiness control over production and marketing of poultry products is in great contrast to the fish component of integrated farming. Agribusiness companies control the breeds, the feeds and the marketing of broiler chickens in central Thailand, and for hen eggs in Northeast thailand (Engle and Skaldany, 1992). In contrast, the fish stocked are purchased from local entrepreneur breeders (Little et al, 1987), fed on poultry and other wastes and marketed directly, or through local middlemen and markets. Farmers are willing to contract-grow broiler chickens over their fishponds for minimal return, in order to gain the benefits of high cultured fish yields. This has resulted in long term declines in the price of both chicken and freshwater fish over the past decade.

Changes in production and demand for poultry and fish stimulate new opportunities for their integration . Increasing proportions of chicken consumed and exported as boneless products has spurred the use of the slaughterhouse wastes as feeds in Thailand, prinicipally for a recently developed hybrid catfish which thrives under such culture conditions (Little et al., 1994). These forms of production however, are concentrated in the hands of relatively few, richer farmers and entrepreneurs. Urban consumers benefit from lower prices for poultry and freshwater fish but rural small-scale production may be constrained through a lack of feeds and markets.

### Environmental and Public Health Aspects

The concentration of nutrients that feedlot agriculture encourages can lead to pollution of surface waters but controlled eutrophication of static water ponds stocked with herbivorous fish can act as on-site treatment. Impacts on surface waters are minimal, providing water exchange is avoided.

Nutrient budgets indicate that although only 15-20 % of input nitrogen and 8-12% of phosphorous are recovered as fish, most of the nutrients accumulate in the sediments (Edwards, 1993). Loss of nutrients with drainage water (<10% of both N and P) (Boyd, 1985). Waste-fed aquaculture is therefore more likely to alleviate pollution from livestock production than cause it, although intensive, long term use of wastes and porous soils could lead to some nutrient losses to surface and groundwater resources.

Public health concerns have been raised about the integration of poultry and fish production on several levels. The risks of direct pathogen transfer to humans in fishponds fertilized with manures, whether consumers, producers or intermediaries, have been most assessed (e.g. AIT, 1986; Buras, 1993). Although faecal bacteria and viruses are present in poultry manures, rapid attentuation of pathogens occurs in most stable, waste-fed ponds (Edwards, 1986, 1993). Clearly, the control of Salmonella and enteric bacteria capable of causing human disease is important. Practically this requires control of these pathogens below threshold levels that can lead to infection of organs and muscle in cultured fish (Buras, 1993).

The control of poultry disease, however, leads to the concern that poultry feeds containing prophylactic antimicrobials can encourages the emergence of antibiotic resistant strains of bacteria. The relative risks are likely to be insignificant compared to other causes such as direct human abuse (Dalsgaard,1993 pers comm.) or chemotherapy of fish themselves (Austin, 1993). There has also been implicit connections made between integrated livestock-fish systems and influenza pandemics (Scholtissek and Naylor, 1988); this disturbing theory has led to widespread comment and discussion of the desirability and impacts of integrated farming (Edwards et al., 1988; Morse,1990; Skladany,1992 ).

The theory maintains that integrated aquaculture encourages the raising of pigs and poultry together to provide manure for fish and that this in turn increases the risks of new forms of influenza developing. Pigs may indeed act as 'mixing vessels' for avian viruses that can transfer to forms more virulent to humans, but fish ponds have had little role in bringing pigs and poultry together on farms. Indeed, pigs, poultry and fish together are rare on both large and small-scale farms in Asia. Intensified poultry-fish systems are more likely to separate poultry from pigs, and other livestock, than traditional farms (Edwards, 1991). The purposeful eutrophication of water leading to blooms of toxic bluegreen algae has also been raised as an issue (Pullin, 1993). The poisioning of mammals drinking water containing toxic strains of Microcystis aeruginosa is established in temperate climates and research has indicated that the Nile tilapia avoids ingesting toxic strains (Beveridge, 1993). Under practical conditions, however, such fish grow fastest in ponds dominated by this same species of algae (Colman and Edwards, 1987). Although the possibility of poisoning of fish and mammals from poultry-waste fertilized water exists, their controlled use in fish ponds reduces the likelihood of pollution to other water bodies

Smallholder mixed crop livestock systems continue to be a dominant agricultural production system in many developing countries, including India. Dairy farming is part and parcel of many such systems, i.e., it is often seen as an important livelihood option to increase household income and to therefore contribute to poverty alleviation in rural areas.

This thesis describes and analyses experiences of BAIF, an Indian NGO, with the use of FSR methodologies in livestock development programmes of Gujarat, India. The objectives were to identify criteria and methodologies for selection of appropriate livestock technologies for farm level, and to identify differences in the methods of selection of appropriate technologies.

Section I describes the variation of livestock production systems in India and gives background to the BAIF organisation and to FSR methodology used. Section II describes the Gujarat research area with agro-ecological-zone-wise information on animal breeds, herd composition, feed resources, crops, and trends in seasonal availability of feed as derived from transects, Participatory Rural Appraisals, and mapping.

Constraint analysis and modelling indicated limited genetic potential of the local breeds and shortage of feed resources, both quantitatively and qualitatively, as major constraints for livestock development. Crossbreeding for breed improvement and use of local feed resources were identified as suitable technologies to alleviate these constraints. Ex-post performance monitoring of BAIF crossbreeding programmes shows that crossbred cattle fitted well in the smallholder mixed farming systems of both tribal and non-tribal farmers in all three selected agro-ecological zones (section III).

Milk production of crossbreds was substantially higher, as was livestock gross margin and household income for households with crossbreds. Various modelling approaches were used in section IV to explore, ex ante, the suitability of feeding technologies such as urea supplementation, use of local and commercial concentrates, and leuceana tree leaves for crop-livestock systems in Gujarat, India. Major conclusions were that (i) concentrate feeding is beneficial to farmers with market access and crossbred cows, (ii) crossbreeding interventions are more remunerative for landless and tribal farmers than for non-tribal farmers; feeding interventions are more effective for crossbreds than for local cows, (iii) maximum farm income is achieved at medium milk yields per animal; higher milk yields require use of better feeds, which renders the straws of the grains useless for feeding; at farm level, the (economically) optimum cropping pattern would then shift from grain crops to cotton.

This section continues with a narrative on BAIF's experiences with field testing of technologies at animal, at herd, at farm and watershed level. Over a period of roughly 30 years, 3 phases in OFTs can be distinguished, i.e. starting with a period of predominantly top-down approaches, moving to a phase with emphasis on participatory identification and testing of technologies, and then into a phase with work at community and watershed level. The studies brought out, among others, that adoption of technologies is facilitated when these use local (feed) resources, that are readily available, require only small changes in farm practices, are relatively simple to implement, and yield tangible results in the short term.

The dynamics in methods and approaches of BAIF's work on livestock development, as it grew from Gandhian roots into a large development organisation. It emphases the dynamics in approaches between top-down, objectivist and reductionist approaches on the one hand and bottom-up constructivist, holistic and self–organised approaches on the other hand. It was concluded that agricultural R&D behaves as a complex adaptive system with its own dynamics and associated paradigm shifts.

The thesis considers development as a continuous process, of which the goals change over time-and-space, Agricultural R&D is ultimately considered to be a complex adaptive system, also in Gujarat, and development organisations like BAIF have to, therefore, show dynamic behaviour. Small ruminants are an important but neglected resource in developing countries. Small ruminant production systems are complex. The multiple goals related to small ruminants, combined with the complexity of their management, and the resources and social arrangements involved, make small ruminants keeping an enterprise that is inherently difficult to study and to understand. This study analysed the behaviour of small ruminant production systems in order to understand their development prospects in different agro-ecological zones in Central Java, Indonesia.

Three districts of Yogyakarta Province, Central Java-Indonesia, were selected to be the research sites, namely Bantul (lowlands), Sleman (middle zone) and Kulon Progo (uplands) districts. The data were gathered by studying literature to obtain primary and secondary data on small ruminant development, interviewing farmers and key persons, group meetings, field observations and monitoring feeding practices, animal performances and marketing strategies, and laboratory analyses on air and water qualities inside and around small ruminant houses. During about one century of small ruminant development in Indonesia, the role of small ruminants has remained more or less the same, whereas major changes occurred in the types of animals kept, in animal numbers and in farmer's management. Driving forces for changes in small ruminant systems have acted at different aggregation levels. The intensification of land use has resulted in major changes in management.

In the middle zone and uplands the majority of small ruminants are kept in confinement, whereas in the lowlands small ruminants are mainly kept in a combination of grazing and confinement. Farmers referred to their small ruminants as a saving (tabungan in Javanese) that provides security and helps to accumulate capital, which in turn helps to reduce hunger and buffers against periodic drought. Manure was the second main reason for keeping small ruminants. The supply and demand of sheep and goats fluctuated throughout the year. The demand for and price of small ruminants increased dramatically during the weeks before Idul Adha, the feast of sacrifice. Farmers do not seem to profit from this increased demand. Farmers rarely sell their animals directly on the small ruminant market or to the consumers. The housing of small ruminants close to the family quarters resulted in very high levels of faecal coliform bacteria and total coliform bacteria, two groups of bacteria used as indicators for water contamination caused by manure.

It is unlikely that small ruminants will become a main income earner in rural households. If households have sufficient family labour for the management of small ruminants, small ruminants are an appreciated secondary activity. Efforts to improve small ruminant production need to be facilitated by stronger institutions, local empowerment and regulation of access to resources. The local government, scientists, extension workers, and farmers themselves have to work together, because improving small ruminant production means that farmers have access to reliable and affordable support services, offering them access to knowledge and inputs, including credit and marketing information.

# Bibliography


Alexander, Paul: *Sri Lankan Fishermen: Rural Capitalism and Peasant Society,* Sterling, New Delhi, 1995.

Bardach, John E.: *Aquaculture: The Farming and Husbandry of Freshwater and Marine Organisms*, Wiley, New York, 1972.

Bartlett, Kim: *The Finest Kind: The Fishermen of Gloucester*, Avon, New York, 1979.

Bryon, Gilbert: *The Lord's Oysters,* Johns Hopkins University Press, Baltimore, 1957.

Carey, George Gibson: *A Faraway Time and Place: Lore of the Eastern Shore*, Arno, New York, 1977.

Carsten, Janet: *The Heat of the Hearth: The Process of Kinship in a Malay Fishing Community*, Oxford University Press, New York, 1997.

Casimir, Michael J., and Aparna Rao: *Mobility and Territoriality: Social and Spatial Boundaries among Foragers, Fishers, Pastoralists and Peripatetics*, Berg, Oxford, England, 1992.

Cecil, Robert Gerald: *Geographical Characteristics of Fisheries in Selected Southeastern Caribbean Island*, , McGill University, 1972.

Cordell, John C.: *Swamp Fishing in Bahia*, Natural History Press, New York, 1978.

Davenport, William H.: *A Comparative Study of Two Jamaican Fishing Communities*, Yale University, 1956.

Davis, Nanciellen: *Ethnicity and Ethnic Group Persistence in an Acadian Village in Maritime Canada*, AMS, New York, 1985.

DePauw, L. W.: *Seafaring Women*, Houghton Mifflin, Boston, 1982.

Ellis, Carolyn: *Fisher Folk: Two Communities on Chesapeake Bay*, University of Kentucky Press, Lexington, 1986.

Erlandson, Jon M.: *Early Hunter-Gatherers of the California Coast*, Plenum, New York, 1994.

Festing, Sally: *Fishermen*, David and Charles, North Pomfret, Vermont, 1977.

Firth, Raymond: *Primitive Polynesian Economy,* Archon, Hamden, Connecticut, 1965.

Fitzhugh, William: *Prehistoric Maritime Adaptions of the Circumpolar Zone*, Mouton, The Hague, 1975.

Frangoudes, K.: *Alternative Management Systems for Fisheries*, Blackwell Science, London, 1999.

Fraser, Thomas M.: *A Malay Fishing Village*, New York: Cornell University Press., 1960.

Gillett, R. D.: *Traditional Tuna Fishing: A Study at Satawal, Central Caroline Islands*, Bishop Museum Press, Honolulu, 1987.

Johannes, R.E.: *Words of the Lagoon: Fishing and Marine Lore in the Palau District of Micronesia*, University of California Press, Berkeley, 1981.

Lewis, Norman: *Voices of the Old Sea*, Viking, New York, 1985.

Malinowski, Bronislaw: *Argonauts of the Western Pacific*, Routledge, London, 1922.

McCay, Bonnie J.: *Oyster Wars and the Public Trust: Property, Law, and Ecology in New Jersey History*, University of Arizona Press, Tucson, 1998.

McGoodwin, James R.: *Crisis in the World's Fisheries: People, problems, and Policies*, Stanford, Stanford University Press, California, 1990.

Peacock, James L., and James C. Sabella: *Sea and Land: Cultural and Biological Adaptations in the Southern Coastal Plain*, University of Georgia Press, Athens, 1988.

Robben, Antonius C. G. M.: *Sons of the Sea Goddess: Practice and Discursive Conflict in Brazil*, Columbia University Press, New York, 1989.

Sather, Clifford: *The Bajan Laut: Adaptation, History, an Fate in a Maritime Fishing Society of South-Eastern Sabah*, Oxford University Press, New York, 1997.

Speck, Frank G.: *Catawba Hunting, Trapping and Fishing*, University Museum, Philadelphia, 1946.

Thompson, Paul Richard: *Living the Fishing*, Routledge and Paul, London, 1983.

Warner, William W.: *Beautiful Swimmers: Watermen, Crabs and the Chesapeake Bay*, Little, Brown, Boston, 1976.

# Index

❑❑❑